The Limitations of Deductivism

Pittsburgh Series in

Philosophy and History

of Science

Series Editors:

Adolf Grünbaum
Larry Laudan
Nicholas Rescher
Wesley C. Salmon

The Limitations of Deductivism

Edited by

Adolf Grünbaum and
Wesley C. Salmon

University of California Press

Berkeley / Los Angeles / London

University of California Press
Berkeley and Los Angeles, California

University of California Press, Ltd.
London, England

Library of Congress Cataloging in Publication Data
The Limitations of deductivism / edited by Adolf
    Grünbaum and Wesley C. Salmon.
        p.   cm. — (Pittsburgh series in philosophy
    and history of science)
    Includes bibliographies.
    ISBN 0-520-06232-9 (alk. paper)
    1. Science—Philosophy. 2. Science—History.
3. Logic.
I. Grünbaum, Adolf. II. Salmon, Wesley C.
III. Series.
Q175.3.L56 1988
501—dc19   88-4824
                                                 CIP

Printed in the United States of America

1 2 3 4 5 6 7 8 9

In Affectionate Memory of
J. Alberto Coffa
(1935–1984)

Contents

Preface   ix
Adolf Grünbaum

In Memoriam: J. Alberto Coffa   xvii
Adolf Grünbaum

In Memoriam: J. Alberto Coffa   xxi
Wesley C. Salmon

Acknowledgments   xxv

Introduction   1
Wesley C. Salmon

1.   Provisos: A Problem Concerning the
Inferential Function of Scientific Theories   19
Carl G. Hempel

2.   Laws, Theories, and Generalizations   37
Ronald N. Giere

3.   Rational Prediction   47
Wesley C. Salmon

4.   The Justification of Deduction in Science   61
Henry E. Kyburg, Jr.

5.   Deductivism Visited and Revisited   95
Wesley C. Salmon

6.   A Nondeductivist Approach to Theoretical
Explanation   128
Frederick Suppe

Contributors   167

Preface

Adolf Grünbaum

In 1980, I organized a Workshop on the Limitations of Deductivism, held on November 7–9 of that year at the University of Pittsburgh, under the auspices of its Center for Philosophy of Science. The workshop's participants were Nancy Cartwright, J. Alberto Coffa, John Earman, Ronald N. Giere, Clark Glymour, Adolf Grünbaum, Carl G. (Peter) Hempel, Henry E. Kyburg, Jr., Larry Laudan, Grover Maxwell, Ernest Nagel, Wesley C. Salmon, Frederick Suppe, and Bas van Fraassen.

The discussion sessions focused on papers given by six of the participants. Some of these authors incurred long delays in the preparation of their essays for the volume of proceedings I had planned. The issues then addressed by the speakers have remained timely. Yet in order to compensate for the lapse of time, some of the contributors—Hempel, Giere, and Suppe—have updated their original presentations, and Salmon ("Deductivism Visited and Revisited") has recently written a new paper for the present volume. It seemed fitting to include as well a second paper by Salmon ("Rational Prediction") because it offers a cogent, concise critique of Karl Popper's deductivist conception of corroboration. Once these essays had all become available, I asked Salmon to join me as coeditor and to write an introduction for the volume. I am indebted to him for having acceded to these requests.

This book was also to include a written version of the informal critical comments that Coffa made at the workshop session devoted to Hempel's paper. Coffa challenged Hempel's thesis that, because of the need for provisos in the explanatory and predictive application of universal laws of nature, even the so-called "deductive-nomological" explanations are not purely deductive. Alas, Coffa's untimely death in December 1984, at the age of forty-nine, aborted the preparation of his thoughtful response to Hempel for publication.

Since this book is affectionately dedicated to Coffa's memory, it seemed appropriate for both Salmon and me to include our reminiscences of him, which we originally presented at his memorial service, held in January 1985 on the campus of Indiana University at Bloomington, where he had been professor of the history and philosophy of science. We also wished to include some account of Coffa's critique of Hempel and its relation to the views of the other contributors to the volume. Portions of Salmon's introduction and the remainder of these prefatory remarks are devoted to that task. Nearly seven years after the workshop, I can reliably report the views Coffa presented there only to the extent allowed by two personal letters I received from him within two weeks after the November 1980 session.

Coffa originally coined the fetching label "deductive chauvinism" for the following conception of the explanatory enterprise: If science is to provide genuinely satisfactory understanding of the world, then the laws of nature must allow deductively formulated accounts of individual events. He did not deny outright that the deductivist might legitimate his chauvinistic regulative ideal. Yet, in his doctoral dissertation,[1] Coffa issued a sobering caveat for the aspiring deductivist: If the lawlike premises of deductive-nomological explanations are not to be patently false, we must explicitly recognize the explanations' auxiliary reliance on qualifying clauses, commonly tacit, which govern such premises. For example, the law that solid rods expand when heated is false, unless we exclude all other influences that would change the length of the rod in a quantitatively different way. Coffa dubbed these modifiers "extremal clauses" (ECs) whereas Hempel referred to them as "provisos." But Coffa saw them as capable of rendering *all* of the unstated *ceteris paribus* conditions (and therefore also spoke of them as "that's all" clauses). Thus, as he wrote to me on November 19, 1980:

> The main difference between Peter [Hempel] and me appears to concern the question whether the "provisoes" pose a problem to deductive arguments in science. In my dissertation, I had argued that they do not; I also argued that most philosophers had overlooked them (Carnap and Hempel are the cases discussed in section 1 of chapter IV, and Canfield and Lehrer—I could have added Stegmüller—on pp. 279ff.), and that the failure to notice their presence had been an obstacle to an appro-

priate generalization of the concept of a deterministic disposition to that of a probabilistic disposition or propensity (I criticize Popper on that score, for example).

More generally, Coffa applied the term "deductive chauvinism" to the view that the higher the probability assigned to an explanandum event, the better the explanation yielding it (private communication). And his further admonition is this: unless we can assume an underlying, hidden determinism, we must conclude that the actual irreducible probabilities exhibited by the stochastic facts are sometimes low. Consequently, they do not accommodate the deductivist's explanatory goals.[2]

According to Giere, the so-called *non*statement (semantic) construal of theories can obviate the need for the provisos required when, as in the standard view, theories are taken to be axiomatized systems of statements. He maintains that, by construing laws as definitions that may or may not apply to actual physical entities, the proponent of the semantic view never needs to formulate provisos. For this reason alone, he tells us, the nonstatement account is more adequate than the received construal, which he rejects. By contrast, Coffa is an advocate of the statement conception. And though he himself called attention, in 1973, to the key role of "extremal clauses" (provisos) in scientific explanation and prediction, he distances himself from Hempel's thesis that these qualifiers undermine the purely deductive status of deductive-nomological (D-N) explanations. Yet Coffa shares Giere's reading of the moral that ought to be drawn *if* Hempel's thesis is taken to be sound, a moral that Hempel himself explicitly *rejects* in his paper herein. Thus, Coffa wrote to me on November 14, 1980:

> To begin with, I would argue that if Peter [Hempel] is right (i.e., if, by and large, we can't formulate extremal clauses ("ECs") in the language of the appropriate theory), then we can't formulate the appropriate (true) laws either in that language. The reason is, quite simply, that the ECs (provisoes) appear in the deductive argument leading to a D-N explanation or prediction not merely as premises but also as parts of the laws. Indeed, the reason why they *must* figure as premises is because they appear as parts of the antecedents of laws (or in equivalent modes of occurrence). At the very least, the laws alone must entail (IC & EC) $\supset 0$, where IC are the initial and boundary conditions, EC the proviso and 0 the observation which we aim to explain or predict. Otherwise, Fred Suppe's point must be granted that in all cases envis-

aged by Peter there is simply no way to apply the D-N model (provisoes or no provisoes). . . .

[I]f Peter is right, the axiomatic view of theories . . . is in deep trouble. If we follow the axiomatizers in trying to construe laws as claims, then if Peter is right, "laws" (i.e., the things so-called by scientists) can only be construed as very false statements. The axiomatizer is stuck with a strange picture of scientific activity: the search for ludicrously false claims which are to be applied in ludicrously unsound (or invalid) arguments which shall be called explanations and predictions. It doesn't look good. . . .

Moreover, whereas if Peter is right, provisoes pose a big problem to the axiomatic view, they pose no problem to the semantic [nonstatement, predicate] view. The reason is, once again, that the semantic view doesn't care whether "laws" are true or false. It can thus take a false law—such as, for all x, if x is a metal and x is heated, then x dilates—and transform it into a predicate T such as "x is a dilating system." No one has to know or be able to *say* that the extremal clause EC is fulfilled. Neither the description of the predicate T nor that of the appropriate structures (heated rods or whatever) require us to be able to say that the temperature change is the only relevant factor in the process under consideration. It just happens to be the case that *if* it is (if the EC is true) the structure will be an instance of the predicate, and if the EC isn't true then (usually) the structure won't be a model.

Well, I sure don't like the consequences of Peter's premiss. As I said at the workshop, I don't think that the premiss is right. But Peter might be right. (He's been right before.)

As Coffa noted in his 1977 paper (see n. 2), the deductivist's explanatory enterprise runs afoul of the low probabilities sometimes countenanced by the stochastic sciences, unless one can successfully introduce an underlying, hidden determinism. But, as we know, some of the deductive chauvinists were not chastened by the seeming indications that microcausality is irreducibly statistical. Indeed, the clarion call for deductive explanation of subatomic phenomena came from the hidden-variables physicists in the face of the scorn heaped upon them by their colleagues. Thus, Wolfgang Pauli dismissed Einstein's clamor for such a theory as an anachronistic pursuit by a conceptually arrested member of the old school. Undaunted by such scorn, Bertrand Russell championed deductivism by declaring himself fundamentally baffled by the *statistical orchestration* of the individually undetermined behavior of supposedly noninteracting microentities. In the chapter on "Determinism" in his book *Religion and Science,* he argued:

What is new in quantum mechanics is not the occurrence of statistical laws, but the suggestion that they are ultimate, instead of being derived from laws governing individual occurrences. This is a very difficult conception—more difficult, I think, than its supporters realize. It has been observed that, of the different things an atom may do, it does each in a certain proportion of cases. But if the single atom is lawless, why should there be this regularity as regards large numbers? There must, one would suppose, be something that makes the rare transitions depend upon some unusual set of circumstances. We may take an analogy, which is really rather close. In a swimming-bath one finds steps which enable a diver to dive from any height that he may prefer. If the steps go up to a great height, the highest will only be chosen by divers of exceptional excellence. If you compare one season with another, there will be a fair degree of regularity in the proportion of divers who choose the different steps; and if there were billions of divers, we may suppose the regularity would be greater. But it is difficult to see why this regularity should exist if the separate divers had no motive for their choice. It would seem as if some men *must* choose the high dives, in order to keep up the right proportion; but that would no longer be pure caprice.

The theory of probability is in a very unsatisfactory state, both logically and mathematically; and I do not believe that there is any alchemy by which it can produce regularity in large numbers out of pure caprice in each single case. . . . This is no more than a suggestion, since the subject is too obscure for dogmatic statements. But if it has any validity, we cannot accept the view that the ultimate regularities in the world have to do with large numbers of cases, and we shall have to suppose that the statistical laws of atomic behaviour are derivative from hitherto undiscovered laws of individual behaviour.[3]

Yet Coffa's notion of a stochastic dispositional property, which is embedded in a propensity conception of probability, contains an answer to Russell's question "But if the single atom is lawless, why should there be this regularity as regards large numbers?" In virtue of being attributable to the *single* case after all, even a stochastic dispositional property demystifies the "alchemy" by which the statistical regularity of frequencies in large numbers is born—as Russell would have it—"out of pure caprice in each single case." Moreover, in Coffa's view, the attribution of such a stochastic property is *not* empty or ad hoc like Molière's famous case of the dormitive powers of opium.

Salmon's response to Russell's challenge consists in the rejection of Russell's dichotomy between completely determined individual cases (that can be subsumed under deterministic laws) and instances of "pure caprice." We have, Salmon claims, strong reason to believe that there are occurrences of a third kind, namely, those that fall under irreducibly statistical laws—laws that, incidentally, may be explained by Coffa's stochastic dispositions. Consequently, Salmon agrees with Hempel in maintaining that it is entirely possible to have *covering law* explanations of indeterministic events.

Nevertheless, Salmon recommends that we repudiate the received *inferential* construal of causal explanation altogether. He sees the statistical sciences as indicating that just this construal is a will-o'-the-wisp. Even those who countenance irreducibly statistical explanations are likely to spare themselves a philosophical wild goose chase, he maintains, by renouncing the inferential conception of the basic explanatory relation—a legacy of inveterate deductivism—in favor of a noninferential one. Hence, to eliminate the residue of deductive chauvinism, he claims, we must reject the notion that probabilistic explanations are *arguments* warranting the inductive expectability of the event to be explained, a notion that is fundamental for Hempel's inductive-statistical (I-S) model of statistical explanation. In short, Salmon offers his concept of degree of probabilistic causal relevance as the basic explanatory relation, in contrast to Hempel's notion of high inductive probability. This relevance relation can readily be assimilated to Coffa's stochastic propensity.

In the same vein, Coffa as well as Suppe see Hempel's I-S schema as an expression of generalized deductive chauvinism, precisely because it is a kind of poor man's forlorn D-N model, embraced *faute de mieux* in a spirit of gratitude for small quasi-deductive mercies, as it were. Indeed, as Salmon points out in "Deductivism Visited and Revisited" (herein), a cognate crypto-deductivism lurks in the minds of those who claim that insofar as individual events are not determined by their antecedents, such occurrences simply cannot be explained at all. Salmon is at pains to discredit the following three versions of this contention: (a) there is no such thing as inductive or statistical explanation; (b) only deductive-statistical explanations are viable, inasmuch as they account deductively for statistical regularities but do not explain individual events; (c) we can explain why single events have certain probabilities but not why they occur. Kyburg's paper challenges yet other important facets of deductivism.

What of Karl Popper's elaborate system of *falsificationist* deductivism? Invoking his own historiography of inductivism, he offers a rival, falsificationist account of scientific rationality in which degrees of falsifiability are claimed to yield (i) a criterion of demarcation between science and nonscience or pseudoscience, (ii) a solution to the problem of induction, (iii) a philosophical theory of corroboration, (iv) an avenue to scientific theories of increasing verisimilitude, and (v) a criterion for the admissibility of auxiliary hypotheses.

Qua species of inferential deductive chauvinism, Popper's falsificationism does, of course, incur the liabilities of its genus. But, clearly, the genus, as such, is not necessarily beset by all of the failings of Popper's falsificationist items (i)–(v) above. In the paper "Rational Prediction," Salmon argues that Popper's purported solution to the problem of induction is quite unsatisfactory. And in a series of four articles of mine published in 1976, which Salmon cites in his introduction (n. 20), the reader will find the details of a gamut of difficulties that I have marshaled against all of the above five prongs of Popper's falsificationism. Since 1976, I have extensively elaborated on my treatment of Popper's item (i), with particular attention to the critique of his thesis that psychoanalytic theory is unfalsifiable.[4]

When Alberto Coffa coined the phrase "deductive chauvinism," he did much more than invent a clever term. He recognized—and focused sharp attention upon—a pervasive and often unconscious tendency to force philosophical concepts and theories into a deductive mold. Whether the tendency is legitimate or illegitimate in one context or another, we need to be aware of what we are doing when we indulge in this practice.

Notes

1. J. Alberto Coffa, "Foundations of Inductive Explanation" (Ph.D. dissertation, University of Pittsburgh, 1973; Ann Arbor, Mich.: University Microfilms, 1973).

2. J. Alberto Coffa, "Probabilities: Reasonable or True?" *Philosophy of Science* 44(1977):186–198.

3. Bertrand Russell, *Religion and Science* (London: Butterworth, 1935), pp. 159–161.

4. Adolf Grünbaum, *The Foundations of Psychoanalysis: A Philosophical Critique* (Berkeley, Los Angeles, London: University of California Press, 1984), chap. 1. Furthermore, see my exchange with Karl Popper in the review-symposium on "Foundations," *The Behavioral and Brain Sciences* 9 (1986):217–284, esp. pp. 220–221, 254–255, and 266–270.

In Memoriam: J. Alberto Coffa

Adolf Grünbaum
January 19, 1985

In the early 1960s, Alberto first wrote to me from Buenos Aires. He said that he had earned a master of science degree in engineering from the university there, and that he was earning his living as an engineer. In remarkably good English, he told me of his strong interest in the philosophy of science. When explaining what particular issues excited him, he unfolded an astonishing mastery of the literature that would have done any fully trained professional proud. Yet he had acquired it autodidactically. The point of his letter was to explore possibilities of coming to this country to earn a doctorate. But his letter was not only impressive; it was positively moving. The more so, perhaps, because—being an immigrant myself—I strongly identified with his aspiration.

In my reply, I encouragingly asked him for any materials that I could use in making a case to secure support for him as a graduate student at the University of Pittsburgh. My memory deserts me now as to why this exploration was not pursued further just then. But I vividly recall my disappointment that, for some reason, he had become reluctant to apply at that time.

It was not until 1966 that I heard from him again. He was on his way to the University of Notre Dame, where he had gotten a fellowship in the Department of Mathematics. By some happenstance, he had a stopover at the Greyhound bus terminal in Pittsburgh. And he phoned me just to say hello. I was so very glad that he had surfaced. But it took all the persuasion I could muster to convince him to let me bring him home to dinner. That evening, it was love at first sight. A few months later, he wrote from South Bend that his mathematics courses were not really his cup of tea.

At the time, Nuel Belnap was director of graduate studies in our department. And to my delight, Nuel agreed with me that Alberto was far more than an excellent bet.

Not long after Alberto's arrival in Pittsburgh, at a meeting of the Greater Pittsburgh Society for the History of Science in November 1968, he gave a paper on "Galileo's Concept of Inertia." This contribution was then published in 1968 in the international journal *Physis*. On the inscribed reprint he gave me, he characterized the paper as "the first not entirely worthless (one) that I publish." This modesty of his often showed itself in a *childlike* surprise when his intellectual accomplishments, or his worth as a person, were being recognized. It was almost as if the esteem of others produced cognitive dissonance in him.

When he fell in love with Linda Wessels, he wrote me a letter of sheer joy, but expressing utter disbelief that such a wonderful thing could ever have happened to *him!* In just this vein he wrote to me in 1974 after we had spoken on the telephone: "As you know, I am not an articulate person, particularly in personal intercourse. I always fear that my spontaneous reaction will be uncalled for and out of place, so I repress it, but I never quite know what to substitute for it. Yet Linda is a witness to the number of times I clicked my heels as soon as I hung up [the phone]." Indeed, his wonderful *joie de vivre* intermittently broke through his shyness, if only in his infectiously hearty, even roaring, laughter. Small wonder that not only I but also Thelma, my wife, became strongly attached to him.

Even before he received his M.A. in philosophy in 1970, he had published four papers in well-known periodicals. In his 1973 doctoral dissertation on "Foundations of Inductive Explanation," which I had the privilege to direct, he presented a highly interesting alternative to the widely discussed models of Carl Hempel and Wesley Salmon. Hempel's inductive-statistical model had relied on the notion of logical probability, and Salmon's statistical relevance model was grounded in a frequency interpretation of probability. Instead, Alberto proposed an account of inductive explanation based on a dispositional version of the propensity interpretation of probability, which had been proposed by Karl Popper.

After he received his Ph.D., he soon branched out into other areas, notably the philosophy and history of geometry, and the philosophical history of logical empiricism. I recall how he held the audience rapt at a Chicago meeting of the Philosophy of Science Association, when he gave a superb paper on the development of the ideas of the Vienna Circle. Indeed, he was an unusually talented, original, penetrating, and thorough scholar, endowed with a creative imagination and a wide-ranging knowledge of his field, as well as of the mathematical physics needed for his intellectual enterprise. That endeavor engagingly blended the history of science *and* of philosophy with keen philosophical analysis. And it was leavened by a subtle sense of humor, which one enjoys as a bonus when reading his work. His substantial remaining writings will be published in a volume entitled *To the Vienna Station*.

I cannot refrain from telling you a little story relating to the *psychological* side of his interest in the philosophy of space-time. Forgive me if in doing so I employ a psychoanalytic idiom. It would seem that the personality types represented among space-time philosophers include two special ones at the opposite ends of the spectrum. The first, *healthy* sort is typified by Wesley Salmon and John Winnie, for example. Its members are excellent geographers, don't lose their way when they drive somewhere, and *their* interest in space-time theories is a wholesome extension of their terrestrial competence. I am, alas, an example of the opposite sort. Long ago, my interest in the field was diagnosed as a massive compensatory neurosis, a vain attempt to cope with my inability to know where I am going. And I cannot deny that I found it comforting to discover in Alberto a soul mate in this psychological respect as well. One evening, he, Thelma, and I had a quiet dinner just by ourselves in the center of Salzburg in Austria. The occasion was the 1983 International Philosophy of Science Congress. After dinner, when it was time to go back to our hotels, he told us that he had rented a car, had parked it in town, and would drop us off at our hotel. But when we walked to where he thought he had parked, there was no car. It became instantly clear from the expression on his face that the hypothesis of car *theft* had no merit whatever. Indeed, he had only the vaguest clue as to the setting in which he had left it. It was dark, and the three of us walked around for easily an hour and a half before we *happened* to come upon his vehicle. When he dropped us off, he said imploringly to Thelma and me, "Please don't tell Linda!" When he was back home, he wrote a letter to Elizabeth McMunn, who is my secretary, and with whom he had had a warm friendship since his days as a graduate student. In his letter to her, he wrote, among other things, "Adolf must have told you that we had a *pretty* good time in Salzburg," lessened only by the car incident.

Two episodes epitomize his character and loyalty. He felt a responsibility toward the Spanish-speaking intellectual world, and hence declined any and all remuneration for work done in that cause, such as preparing Spanish translations of philosophical writings in English. Thus he refused a handsome fee offered to him for services he rendered to the philosophical community in Mexico City. It was heartwarming to see his response to the replacement of the brutal military dictatorship in Argentina, and also his joy when he received a high-level invitation to attend Raul Alfonsin's presidential inauguration there. When he taught in Pittsburgh in 1976, he selflessly devoted many extra hours of tutoring, far beyond the call of duty, to fill lacunae in the technical preparation of his graduate students.

Alberto was a gem of humanity. His death leaves a gaping hole in my heart, a wound that will never heal.

In Memoriam: J. Alberto Coffa

Wesley C. Salmon
January 19, 1985

Others are speaking today of Alberto's personal qualities as a man—his wit and vitality, his capacity for deep feelings, his generosity, his integrity. Like the rest of you, I appreciate intensely such crucial human characteristics, and honor Alberto for them. However, my remarks will focus on his superb philosophical and intellectual skills, which represent an extremely significant aspect of his life. I shall talk about his quality of mind.

Several weeks ago, while putting some papers in order in my office, I came across a reprint of one of Alberto's articles.[1] Its title sounded interesting and it was one I had not read, so I decided to read it before filing it away. It dealt with some of Russell's work on the foundations of mathematics, in particular, the effort to furnish an account of mathematical demonstration that would eliminate appeals to intuition as elements of proofs. Not only did it turn out to be interesting—it was exciting. Alberto had recounted a situation in turn-of-the-century mathematics that in certain ways strongly paralleled one in contemporary inductive logic. It helped me to see the problem in a novel way. I looked forward to seeing Alberto and talking about it with him. Alas, it can never be.

I relate this very private incident to make a point about Alberto and his way of philosophizing. He was, in a word, a *profound* thinker—one of the few genuinely profound philosophers I have ever known. His profundity was never the sham variety that deals in obscurantism or pretension. It was a genuine profundity that cut to the core of a problem and brought out what is essential in terms of the simple fundamentals. This basic intellectual trait was evident when I first knew him—he a graduate student at Pittsburgh and I a visiting faculty member. During that year (1968–69) Alberto was choosing a dissertation topic, and as a courtesy to a visitor, I was on his committee. I still recall quite vividly his presentation of his prospectus. He began by spelling out the fundamentals, item by item, in a numbered list. I was on the verge—and I believe Adolf was too—of becoming impatient with the rehearsal of points that we considered familiar to all of us. Then Alberto began drawing startling conclusions that were not a bit familiar. Alberto saw implications of the fundamentals in ways that none of us had noticed before. It was simply a case of incredibly clear philosophical thought—a matter of spelling out the premises clearly and explicitly, and then drawing the inevitable conclusions—a gift so few of us have.

There is, I suspect, a strong analogy between such profound philosophical reasoning and the kind of thing that is sometimes found in pure mathematics or high-level theoretical science—when an extremely general result is established by a simple and elegant demonstration. I think immediately of such examples as Cantor's proof that the subsets of a set outnumber its members, without restriction, whether the given set is null, finite, or infinite. Another case in point: the ancient demonstration that among all of the rational numbers there is none whose square is two.

Please forgive me for talking shop; I ask your indulgence. The intellectual life was a major component of Alberto's life. The examples to which I am referring are the *beautiful* results; their aesthetic value is enormous to those who take pleasure in contemplating such things. Forgive me also for choosing cases in which I was directly involved, for those are the cases in which I can most clearly see his intellectual elegance and power. The specialist can best appreciate the profundity, simplicity, and loveliness of such achievements.

Let me mention just one additional beautiful example of Alberto's work that touched me directly. Alberto's dissertation dealt with scientific explanation—statistical or probabilistic explanation in particular.[2] It was a topic on which I had been working for several years. We began discussing it during my visit to Pittsburgh, and continued subsequently when we were colleagues here in Indiana. One day he gave me a copy of a paper he had written entitled "Hempel's Ambiguity." It was not an attempt to fault Hempel's work on the basis of Hempel's ambiguous use of terms; rather, it was an analysis of Hempel's doctrine of the fundamental ambiguity of inductive-statistical explanations and the epistemic relativization that results. I read it with great excitement; it struck me then (and still does) as the most penetrating critique of Hempel's theory that had come to my attention. So great was the excitement that I was moved immediately to write a follow-up note drawing some conclusions that were pertinent to my viewpoint. Our articles were published together in *Synthese*.[3] In analyzing the very concept of epistemic relativization, crucial to Hempel's account, Alberto saw that Hempel's treatment deprives inductive-statistical explanation of any independent status, rendering it parasitic upon deductive-nomological explanation. I realize that these remarks may not mean much to anyone unfamiliar with the philosophical issues involved, but I hope they give everyone some sense of the depth of Alberto's intellectual contributions.

Alberto's mind had another aspect, quite closely related to his profundity, I think. It was the historical dimension. It seemed that he felt we cannot appreciate where we are philosophically unless we know where we came from. Thus, Alberto devoted an enormous amount of time and energy to the study of such towering figures in twentieth-century scientific philosophy as Carnap, Schlick, Reichenbach, Russell, and Wittgenstein. In recent years I found particular delight in the many historical tidbits he would mention whenever we met. In our last conversation he told me of a postcard he had discovered in Reichenbach's correspondence strongly suggesting that Reichenbach's well-known doctrine of the conventionality of geometry—one of the things for which he was most famous—may well have been a gift from Schlick rather than his own original contribution.

Alberto's determination to get the history right was, I believe, a manifestation of his fierce intellectual honesty. It is regrettable that much of his recent historical work has not, as yet, been published.[4] However, some of it has been published. Speaking again on matters that touched me, I must comment on his substantial "Explanation and Clarification" of Reichenbach's *Experience and Prediction,* which appears in German in the definitive nine-volume Vieweg edition of Reichenbach's *Collected Works*.[5] It puts in perspective the place of Reichenbach's major epistemological treatise (originally published in English and never previously published in German translation) in the development of logical empiricism. Another of his published works that is special to me is his beautiful discussion of the controversy between Reichenbach and Weyl on the foundations of relativity theory.[6]

I could go on at length giving further examples of Alberto's intellectual achievements, but I won't. I'm quite overcome by our personal and intellectual loss.

We all affectionately remember Alberto as a person of deep feelings, of great generosity, and complete integrity. It is, consequently, no surprise that his intellectual work has reflected these same virtues. The philosopher we admire is not so very different from the man we love.

Notes

1. J. Albert Coffa, "Russell and Kant, *Synthese* 46 (1981):247–263.

2. J. Albert Coffa, "Foundations of Inductive Explanation" (Ph.D. dissertation, University of Pittsburgh, 1973; Ann Arbor, Mich.: University Microfilms, 1973).

3. J. Albert Coffa, "Hempel's Ambiguity," and Wesley C. Salmon, "Comments on 'Hempel's Ambiguity' by J. Alberto Coffa," both in *Synthese* 28 (1974):141–164 and 165–174.

4. Though, as Adolf Grünbaum mentions in his eulogy, a substantial portion of it will be published by Cambridge University Press under the title *To the Vienna Station*.

5. J. Alberto Coffa, "Erläuterungen, Bermerkungen, & Verweise zu dem Buch *Erfahrung und Prognose*," in Andreas Kamlah and Maria Reichenbach, eds., *Hans Reichenbach, Gesammelte Werke*, Band 4 (Braunschweig and Wiesbaden: Vieweg, 1983), pp. 255–297.

6. J. Albert Coffa, "Elective Affinities: Weyl and Reichenbach," in Wesley C. Salmon, ed., *Hans Reichenbach: Logical Empiricist* (Dordrecht: Reidel, 1979), pp. 267–304.

Acknowledgments

The editors wish to express their gratitude to:

The R. K. Mellon Foundation and the Sarah Mellon Scaife Foundation for financial support of the Workshop on the Limitations of Deductivism, held at the University of Pittsburgh in 1980.

The editors of *The British Journal for the Philosophy of Science* for permission to reprint "Rational Prediction," by Wesley C. Salmon.

Linda Wessels, widow of J. Alberto Coffa, for information concerning the contents of his literary *Nachlass*.

Introduction

Wesley C. Salmon

The roots of deductivism go back to antiquity—to the very dawn of philosophy. Among the gifts bequeathed to us from the ancient Greeks, two stand out as particularly important. The first of these is Euclidean geometry. If the traditional stories are to be believed, Thales of Miletus—who flourished around 600 B.C., and is regarded by many as the first philosopher—visited Egypt, where geometry was employed as a practical art. He brought it back to Greece, and transformed it into the beginnings of a deductive science. He is said to have introduced the notion of *proof,* and to have provided proofs of several theorems. About three centuries later Euclid wrote *The Elements,* in which geometry is presented as an axiomatized deductive system—a system in which a large number of propositions (theorems) are demonstrated on the basis of a very few assumptions (axioms or postulates). Until Newton's *Principia* (published in 1687, just 300 years ago) no other body of scientific knowledge could begin to match Euclidean geometry in rigor, scope, and fecundity. For about two millennia Euclidean geometry was *the paradigm* of scientific knowledge. This fact has had a powerful influence on the history of philosophy from antiquity to the present. Among other things it suggested that the *only* proper form for a respectable scientific theory is a deductive axiomatic system.

The second great gift from antiquity is deductive logic, comprising the syllogistic logic of Aristotle and the propositional logic of the Stoics. These systems of logic, which were not significantly improved or extended until the nineteenth century, provided analyses of the nature of deductive proof. Although, even taken together, they are not nearly strong enough to account for geometric reasoning, they did provide considerable understanding of the nature of deductive validity. They showed what logical demonstration

1

amounts to—how the truth of the premises of an argument could guarantee the truth of its conclusion.

During the nineteenth century the non-Euclidean geometries were discovered, and the foundations of geometry and those of the infinitesimal calculus were purified. These important developments, which ushered in modern mathematics and modern logic, did nothing to undermine the deductive sciences. Instead, they improved the power and rigor of deductive methods. They showed that deductive demonstration had resources not dreamt of in antiquity.

The present collection of six essays, *The Limitations of Deductivism*, is in no sense an attack on deductive methods as they are practiced in modern mathematics and logic. In these so-called *formal sciences* the use of deduction is unchallenged. Indeed, all of the authors here represented rely heavily upon deductive techniques in their own work. At issue is the question of the extent to which deductive logic is adequate to characterize what goes on in the empirical sciences—the physical, biological, and behavioral sciences. *Deductivism* (read: *deductive chauvinism*) is the view that the *only* logical devices required in the empirical sciences are deductive.

It was our admired and beloved colleague, J. Alberto Coffa, to whom this volume is dedicated, who coined the apt term "deductive chauvinism" to characterize the view that deduction does it all. His main application of this concept was in the context of scientific explanation, but we can—without violating his intent, I think—distinguish two forms of deductive chauvinism, both of which are championed by Karl R. Popper and his followers.[1] The first is the view that the only legitimate arguments are valid deductions—that is, neither induction nor any other form of nondemonstrative inference has any proper role in science. It could appropriately be called *inferential deductive chauvinism*. The second is a more widely held view—advocated by many philosophers, including, for example, Wolfgang Stegmüller[2] and G. H. von Wright[3]—that the only admissible scientific explanations are deductive-nomological. This version might be called *explanatory deductive chauvinism*. If all acceptable explanations are deductive-nomological, explanatory chauvinism is a special case of inferential chauvinism. However, one can be an explanatory chauvinist without being an inferential chauvinist, for one can maintain that there are nondeductive arguments that are not explanations. Likewise, one can be an inferential chauvinist without being an explanatory chauvinist, for one can deny that explanations are arguments of any sort.

The papers in this volume look at deductivism from three directions. First, Carl G. Hempel, Ronald N. Giere, Wesley C. Salmon ("Rational Prediction"), and Henry E. Kyburg discuss the nature of scientific inference; they are dealing with inferential deductive chauvinism. Second, Salmon ("Deductivism Visited and Revisited") and Frederick Suppe deal with the nature of scientific explanation; they are concerned with explanatory deductive chauvinism. Third, Giere and Suppe, in their efforts to deal, respectively, with scientific inference and scientific explanation, address the nature of scientific theories. They contrast what has come to be known as *the standard view of theories* with the more recent *semantic view of theories*. Giere employs the semantic conception to defend deductivism against a difficulty raised by Hempel. Suppe appeals to the semantic conception to develop a nondeductivist account of scientific explanation.

Inasmuch as an understanding of the nature of scientific theories is central to the whole discussion, I shall begin with that issue.

Theories

One strong version of deductivism—let us call it *narrow deductivism*—identifies scientific theories as axiomatic systems, closely analogous to Euclidean geometry in its modern rigorous form. From a clearly delineated set of axioms one deduces the theorems that capture the content of the domain to which that particular axiomatic system is intended to refer. Before the advent of non-Euclidean geometries, the axioms of Euclidean geometry were widely regarded as self-evident truths; we now understand that, from the standpoint of the axiomatic system, they are simply assumptions from which we derive consequences. Their truth or falsity is not at issue.

Indeed, the study of the different geometries and other axiomatic systems led to the conception of a formal system, and to the distinction between interpreted and uninterpreted formal systems. For instance, we may look at a geometry as an uninterpreted system. It consists of primitive "propositions" and derived "propositions." These "propositions" contain a variety of terms. Some terms—such as "and," "not," "if," "all," and "only"—have well-established meanings in logic; others—such as "one," "two," "equal," and "add"—have well-established meanings in arithmetic. Terms of these types have no special geometrical meanings. Other kinds of terms are characteristically geometrical—for example, "point," "straight line," "parallel," "intersect," "circle," "triangle," and "congruent." Some of these are definable in terms of the others. If, for instance, we understand "straight line" and "intersect" we can obviously define "parallel." But not all of these terms can be defined in terms of the others; some must be taken as primitive. So we have two classes of terms: primitive and defined. With respect to the axiomatic system as such, the primitive terms have no meanings whatever. And since the defined terms are defined by means of the primitive terms, the defined terms have no meanings either. That was my reason for putting the word "proposition" in scare quotes; given the fact that the primitive and defined terms that appear in them do not have meanings, neither do the axioms and theorems have any meaning. A system of this sort is an *uninterpreted axiomatic system.*[4]

An uninterpreted axiomatic system becomes an interpreted system if meanings are assigned to the primitive terms. Once the primitive terms have been interpreted, the defined terms inherit meaning from them. When both the primitive and defined terms have meanings, the axioms and theorems become meaningful propositions (no scare quotes). Given that these propositions are meaningful, they are either true or false. If the axioms turn out to be true under a given interpretation, the theorems, on the same interpretation, will also be true. This is what is meant by saying that the theorems follow deductively from the axioms: on any interpretation that renders the axioms true the theorems will be true as well.[5]

From the standpoint of pure mathematics or pure logic, it may be perfectly proper to deal with uninterpreted systems alone, though I doubt that mathematicians or logicians actually work that way very often. Be that as it may, the empirical sciences are concerned with knowledge of the real world. Uninterpreted axiomatic systems, as such, have no connection whatever with empirical phenomena. If these axiomatic systems are to have any connection with reality, they must be interpreted. When an interpretation has been given, it becomes a matter of empirical investigation to find out whether the axioms, and consequently the theorems, are true.

Immanuel Kant, who died at the beginning of the nineteenth century, had a profound influence on nineteenth- and twentieth-century philosophy. Geometry played a key role in his philosophy, inasmuch as he maintained that we can have synthetic a priori knowledge of the structure of space. The discovery of non-Euclidean geometries early in the nineteenth century, and the advent of relativity theory early in the twentieth century, also placed geometry in a key role. Thus, for the most influential philosophers of science in the first part of the twentieth century—such as Rudolf Carnap, Hans Reichenbach, and Moritz Schlick—it was altogether natural to focus attention upon geometry, and to emphasize the importance of the distinction between pure and applied geometry.

Against this background there arose what has come to be called the *standard view* of scientific theories. According to this view, a scientific theory is an axiomatic system. This system is given empirical meaning by means of *rules of correspondence, coordinating definitions,* or (as Hempel refers to them in his paper) *interpretative statements* that provide a connection between the terms in the axiomatic system and the entities and phenomena of the world. To provide a theory of physical space, for example, one could say, roughly speaking, that the paths of light rays are straight lines, that standard measuring instruments such as meter sticks provide relations of congruence, and that the intersection of cross hairs in a surveyor's transit defines a point. Given such meanings, it becomes an empirical matter to determine whether Euclidean or non-Euclidean geometry describes the structure of physical space.

From the standpoint of the standard view, it would be nice to think that each primitive term in an axiomatized theory could be given a complete coordinating definition in terms of observables. As Carnap, Hempel, and many other philosophers recognized, this happy situation is seldom, if ever, realized. Upholders of the standard view introduced the notion of a partial interpretation, and a variety of techniques were proposed for achieving partial interpretations. Over a period of several decades, Hempel has contributed importantly to the elaboration and refinement of the standard view of theories, including the concept of partial interpretation.[6] His paper in this volume provides an excellent brief account.

To the thesis of narrow deductivism—construed as the claim that scientific theories must be regarded as partially interpreted formal systems—two of the authors in this volume, Giere and Suppe, take strong exception. They are representatives of a group of prominent philosophers—including Philip Kitcher, Joseph Sneed, Wolfgang Stegmüller, Patrick Suppes, and Bas van Fraassen as well—who adopt what has come to be known as the *semantic view of theories.* According to this view, a theory is not a set of propositions or statements but rather a set of models. Models are simply interpretations, so instead of identifying the theory with an axiomatic system, one identifies it with the interpretations of the system. One thing this means is that two different axiomatic systems that are semantically equivalent—that is, that share the same set of models—constitute one theory rather than two. A theory correctly describes some portion of the world if one of its models is identical with that part of the world; it gives a nearly correct description of any part of the world that is approximately identical with one of the models.

It is natural at this point, I believe, to ask in what fashion or to what extent the proponents of the semantic view are actually rejecting deductivism. If, as I have suggested, deductivism identifies scientific theories with axiomatic systems, it raises the question, what logical resources are available for the derivation of theorems from axioms? This amounts to the question, what precisely is deductive logic? There is some temptation to identify it with first-order logic, since such systems are demonstrably consistent and complete. A theory would then consist of a set of statements—such as "All ravens are black"—that can be formulated in a first-order language.[7] Some of these would be singled out as axioms, and the theory would encompass the axioms and all of the statements that follow from them by the standard rules of first-order logic.

First-order logic is, however, extraordinarily weak. The deductivist—if ever there was such—who maintains that all legitimate scientific theories can be formulated as axiomatic systems in first-order logic has an untenable position. Since first-order logic cannot accommodate any mathematics beyond the most rudimentary, it cannot be an adequate language for any reasonably well-developed science. Deductivism defined in this extremely narrow fashion is surely unacceptable. It is reminiscent of the pre-nineteenth-century situation, mentioned above, in which Aristotelian syllogistic logic was taken to encompass all of deduction and in which Euclidean geometry was the paradigm of a deductive science, in spite of the patent inadequacy of Aristotelian logic to account for geometrical derivations.

It seems clear, then, that any tenable version of narrow deductivism will identify scientific theories with axiomatic systems that use mechanisms more powerful than those of first-order logic. It seems to me that there is nothing inimical to deductivism in using higher-order logic, set theory, or any other branch of mathematics that is suitable and efficient. A word of caution is, however, in order. Just because something is called "mathematics" does not automatically qualify it as a legitimate deductive tool; we must not forget the lesson painfully learned in the nineteenth century regarding the foundations of the infinitesimal calculus. The best efforts of Cantor, Cauchy, Dedekind, and Weierstrass were required to clean up that mess.[8]

Giere and Suppe both reject narrow deductivism even in its less restrictive form. Their appeal to the semantic view of theories calls immediate attention to the basic distinction between syntax and semantics. Syntax deals only with relationships among symbols, without any consideration of the objects to which the symbols are supposed to refer. It is possible to define many basic logical concepts—including the relations of *deductive consequence* and *logical equivalence*—purely syntactically. For example, we can say, roughly speaking, that *a derivation of a theorem from the axioms* is a sequence of uninterpreted formulas, each of which is an axiom or follows from preceding members of the sequence by the formal rules of transformation of the system. The final formula in the sequence, the theorem, is a logical consequence of the axioms of the system. Syntactic considerations of this sort fall within *proof theory*.

Semantics deals with the relationships between symbols and their referents; meaning and truth are semantic concepts. The concept of a model is also a semantic concept. Many of the logical concepts that can be defined syntactically—including deductive consequence and logical equivalence—can also be defined in semantic terms. For example, one formula is a *logical consequence* of another if, on every interpretation that makes the one true, the other is true as well.[9] Considerations of this sort fall within *model theory*. Giere characterizes the difference between the semantic approach and the standard (logical empiricist) approach in terms of the contrast between the methods of model theory and those of proof theory.

If we characterize a scientific theory as an axiomatic system, then we can use that system to determine a set of models; that is, we can start with axiomatic systems and use them to characterize sets of models. We can also look at different axiomatic systems that are semantically equivalent to one another, and point out the equivalence. Alternatively, we can look at a set of models, and seek axiomatic systems that have these models as interpretations; that is, we can start with the models and use them to characterize axiomatic systems. Whichever way we start, we shall certainly be concerned with deductive consequences. Deriving theorems from axioms helps us to characterize and identify the models of that axiom set. That kind of information can be valuable whichever way you go—either from axioms to models or from models to axioms.

It would seem natural to incorporate within the scope of deductive logic the semantical relationships as well as the syntactical ones. In deductive logic—especially as it is applied to empirical subject matter—we are regularly concerned with such semantical notions as the *truth-preserving* character of deductive derivations. And if one examines standard texts in model theory one finds that the techniques are wholly deductive. Model theory does not appeal to inductive inferences or sensory observation. With these considerations in mind, we may define *deductivism in the broad sense* as the thesis that deductive techniques, including those of model theory, are sufficient to analyze the logic of the empirical sciences.

I find it hard to see among the proponents of the semantic conception of theories a group that rejects or undermines deductivism, construed—as I have recommended above—in a reasonably broad fashion. Inasmuch as Giere does not explicitly dissociate himself from deductivism, it appears that he shares this view. When Suppe explicitly rejects deductivism, he seems to have in mind the view that scientific explanations are deductive arguments—indeed, that they are *either* deductive *or* inductive arguments. Suppe sees in Hempel's inductive-statistical (I-S) model of explanation a manifestation of deductive chauvinism, just as Coffa did.[10] It is the attitude that, if we *must* put up with statistical explanations, we should *at least* make them resemble deductive explanations as closely as possible. As a consquence, statistical explanations must be inductive *arguments,* and their associated probability values must be high. Thus, I think we might say, both Coffa and Suppe reject the *inferential conception* of scientific explanation, a point with which I emphatically agree.[11]

What the proponents of the semantic view of theories have demonstrated clearly, I think, is that the standard view of theories is by no means a consequence of deductivism. By freely using set theory and model theory, they have made a convincing case that the resources available to the deductivist for purposes of studying the structure of theories are much more extensive and powerful than was realized by the earlier logical positivists and logical empiricists. It remains to be seen whether the deductivist, even in the broadest sense, can provide adequate accounts of scientific inference and scientific explanation.

Scientific Inference

Whether one adopts the standard view of theories or some other alternative, such as the semantic view, the problem of establishing a connection between a theory and that part of the universe to which it is intended to apply will surely arise. Hempel discusses this issue in terms of interpretative statements. In his paper on provisos in this volume he invites us to consider a situation in which a theory is employed to provide a connection between two distinct observable facts—as when one predicts a future event on the basis of a previously observed fact by means of that theory.[12] In such cases three steps are involved: (1) establishing a connection between the previously observed facts and the theory, (2) making an inference within the theory, and (3) establishing a connection between the theory and the predicted observable fact.

Using an example involving the behavior of bar magnets, Hempel attacks the strict deductivist view by arguing that *none* of the three inferences can be considered deductively valid. His argument goes roughly as follows.[13] Suppose we observe a metal bar to which iron filings adhere. We predict that, if it is cut in two, and the pieces are hung close to one another at the same level by fine threads, they will orient themselves in a straight line. (1) To go from the prior observed facts to the theory is not deductive; it involves *inductive or theoretical ascent*. From the fact that iron filings stick to a bar we cannot deduce that it is a magnet; the iron filings may adhere to the bar because it has a sticky surface. (2) The theoretical inference is not deductively valid because it presupposes various provisos. Our theory tells us that a single bar magnet, cut in two, becomes two bar magnets. But this will not be true if the cutting is done at too high a temperature, for under that circumstance the magnetization will disappear. (3) The transition from the theory to the prediction is nondeductive for the same reason—it, too, requires provisos. If the pieces are hung in a strong magnetic field they will line up with the field and not with one another. Hempel argues that the provisos cannot be incorporated explicitly into the theory, and he maintains that the use of probabilistic connections would not enable us to escape the problem. He offers a fundamental challenge to those who would maintain that the inferences involved in establishing theoretical connections among observable states of affairs—either for prediction or for explanation—must all be strictly deductive.

In an attempt to show that the problem of provisos can be resolved by appealing to the semantic view of theories, Giere introduces another example, namely, a simple pendulum.[14] He discusses the case in which the bob is made of iron, and a strong magnet is in the vicinity. He maintains that suitable definitions of "classical pendulum" and "magnetically augmented pendulum" can be used to take care of the proviso about the absence of magnets; other definitions would, presumably, handle other provisos.[15] Hempel claims, however, that a switch from the standard view of theories to the semantic view would not solve the problem, Giere's efforts notwithstanding. I shall not attempt to adjudicate this dispute.

Hempel's discussion of the limitations of deduction in empirical science makes reference to theories to which we are willing to appeal for purposes of prediction or explanation. He does not raise the question of how such theories are to be established. Two options may *appear* to be open to the deductivist. The first is Popper's deductivism; the second is the traditional hypothetico-deductive method. I shall suggest, however, that this second alternative is not available to the deductivist because it is inherently inductive.

According to Popper, we test theories by deducing, with the aid of observed initial conditions and auxiliary assumptions, some consequence whose truth or falsity can be checked by observation. If we ascertain, by observation, that the prediction was false,[16] we can apply *modus tollens* to conclude that something among the premises is false. This is a deductively valid inference. If, moreover, we are unwilling to reject either the statement of initial conditions or the auxiliary assumptions, we must conclude that the theory being tested is false.[17] Popper frequently refers to his approach as the method of *conjectures and refutations*. We conjecture bold hypotheses for testing; the only test outcome that decides the fate of the hypothesis is the negative one that refutes it. We never confirm hypotheses on the basis of positive evidence, or in any way render one unrefuted hypothesis more credible than another. We only reject hypotheses on the basis of negative evidence.

My chief argument against Popper's inferential deductivism hinges on the predictive function of scientific theories. If we need to appeal to a scientific theory for purposes of prediction or rational planning, we need to know how to select the appropriate theory. On Popper's view we can only reject hypotheses that have been refuted, we cannot establish any—even tentatively—on the basis of positive evidence. Given this restriction, no matter how much empirical evidence we have accumulated, we are always left with an (in principle) infinite collection of unrefuted theories. How are we to choose among them?

Popper has introduced the concept of *corroboration*, which is not to be confused with the hypothetico-deductive theorist's concept of confirmation. The corroboration rating of a theory is simply a *record of its past performance* in passing severe tests without being refuted. The record of past performance cannot, however, provide any guide whatever to the future happenings that determine the success or failure of our practical ventures. For that, some kind of inductive reasoning is required. The basic argument supporting this conclusion is quite simple. The observations we have made so far provide us with information about past and present facts, but not about future facts.[18] Deductive arguments are nonampliative; the conclusion of a valid argument has no content that is not present in its premises. If observation and deduction are our only resources, we cannot legitimately draw any conclusions—even probabilistic ones—about the future. The Popperian approach cannot provide the predictive information we need for rational planning.[19]

According to the traditional hypothetico-deductive method, the negative outcome has the result Popper maintains. If, however, the outcome of the observational prediction is positive, we may claim that the hypothesis has been confirmed to some extent by that result. It should be noted most emphatically that the hypothetico-deductive method of *confirmation* is not deduction. If we claim that it is a valid deductive argument we are guilty of the fallacy of affirming the consequent. The only way to save the hypothetico-deductive method is to claim that it is inductive in character. The hypothetico-deductive method is deductive to the extent that it requires the deduction of an observational prediction from the hypothesis, initial conditions, and auxiliary assumptions. When it passes judgment on the hypothesis on the basis of positive test results it has left the realm of deductive inference.

Whether the hypothetico-deductive schema is to be used only negatively (as in Popper's account) or positively (as in the traditional hypothetico-deductive account), Kyburg finds a serious difficulty from the outset. If one is to use that schema to pass any kind of judgment upon hypotheses, we must have statements of initial conditions that we are prepared to assert on the basis of observation. In many realms of science we are concerned with quantitative hypotheses, and in these areas the establishment of both initial conditions and the correctness or incorrectness of the observational prediction requires measurement. As a result of a careful analysis of what is involved in measurement, Kyburg concludes that we are seldom, if ever, justified in claiming that the quantitative results that issue from measurement are true. The theory of errors enables us to make probabilistic statements about the actual values of the quantities that have been measured, but not an unqualified statement that can be used as a premise in a deductive argument. There is, he maintains, a layer of probabilities between observation and theories. In order to get around the difficulties he sees in deductivism, Kyburg proposes adoption of a fairly complex inductive model, some of whose features he spells out in his paper herein.

It is my conviction that inferential deductive chauvinism has suffered defeat by the essays in this volume in three ways. The first is Hempel's argument concerning provisos and theoretical ascent. The second is Kyburg's point about measurement and the treatment of errors. The third is my argument to the effect that pure Popperian deductivism cannot provide the kind of guide to the future that is needed for rational behavior.[20]

Scientific Explanation

Although Hempel has done more than anyone else to articulate, elaborate, and defend the deductive-nomological (D-N) model of scientific explanation, he *never* said that all legitimate scientific explanations are of the D-N type. Even in the classic 1948 paper that he wrote with Paul Oppenheim it is explicitly acknowledged that some legitimate scientific explanations are of a statistical or probabilistic sort.[21] In subsequent writings he went on to spell out the details by elaborating the deductive-statistical (D-S) and inductive-statistical (I-S) models. Since the D-S model characterizes a type of statistical explanation as a deductive argument, its existence raises no problem for the deductivist. It is fair to say, I believe, that Hempel's theory of scientific explanation, incorporating the foregoing three models of explanation, constituted the *received view*, at least until fairly recently.

The I-S model has experienced many vicissitudes since it appeared on the scene—serious enough, in my opinion, to merit its rejection. Suppe is correct in remarking that the I-S model is inspired by deductivism and rests upon deductivist intuitions. This point was brought out most forcefully in two papers by Coffa.[22] My response was to offer the statistical-relevance (S-R) model as a replacement for it. This model is doubly offensive to deductivists; according to it scientific explanations are not only nondeductive, they are not even arguments of any sort. The S-R model has suffered its own vicissitudes. Various philosophers have rejected the S-R model for various reasons; I rejected it in favor of a causal model, in which it must be understood that causality is construed broadly enough to encompass probabilistic causality.

Considering all of the difficulties that have surrounded the notion of probabilistic or statistical explanation, it is tempting to retrench, and to maintain that all legitimate scientific explanations are deductive—they are either D-N or D-S. In fact, there is no good reason of which I am aware to separate those two types. From the beginning Hempel has claimed that there are D-N explanations of general laws as well as D-N explanations of particular facts. D-S explanations are also deductive explanations of general laws, the laws in question being statistical rather than universal. Nevertheless, a D-S explanation of a statistical law is a deductive argument having that law as its conclusion and containing at least one law among its premises. The law in the explanans will normally be a statistical law, but that consideration does not, I think, merit making a separate model for D-S explanations.

There is an important caveat to be entered here. The 1948 Hempel-Oppenheim paper provides an explication of D-N explanation of particular facts but not of general laws. In the notorious footnote 33, the authors call attention to a serious problem in connection with explanations of laws, and they confess to having no solution. The problem is this. There are bona fide explanations of laws, such as the derivation of the law of conservation of momentum from Newton's second and third laws of motion. We can also cook up patently spurious "explanations," such as the derivation of the law of conservation of momentum from the conjunction of that law and any arbitrary unrelated law, such as the law of diminishing marginal utility of money. How can we rigorously characterize the difference between these two types? Hempel did not return to this question to provide an answer. An attempt by Michael Friedman in 1974 was not successful.[23] As things now stand, the deductivist does not have any account of explanations of laws. The promissory note that has been given, at least tacitly, regarding explanations of laws is beginning to look like an enormous third-world debt on which a bankrupt nation has defaulted. Nevertheless, the nondeductivist should not feel smug, for I do not think anyone else has a satisfactory account either.

In my paper "Deductivism Visited and Revisited" (chapter 5 in this volume), I try to give explanatory deductivism the best run possible for the money. In the end, however, I find two recalcitrant problems. First, there are situations in which causal considerations clash with explanatory considerations if we adopt the D-N model. It seems clear that in some cases, at least, we adopt a sine qua non concept of causality. In these cases the cause is a necessary condition, whereas the D-N model demands a sufficient condition for explanation. Thus, counterintuitively, we find some examples in which—according to the deductivist—there is a cause that does not explain and others in which there is an explanation that does not include the cause.[24] Second, however strongly we argue that pure science can dispense with probabilistic explanations of particular facts, applied science cannot. In applied contexts we are not satisfied with D-S explanations that allow us to deduce probability distributions for unlimited classes of cases; we seek *and find* probabilistic explanations of particular occurrences, such as airplane crashes, bridge collapses, and fires. I conclude with the thought that a philosophy of science that confines itself to pure science, totally neglecting applied science, is quite hopelessly biased.

Suppe finds serious shortcomings in the models of explanation Hempel has offered and in the alternatives with which I have suggested replacing them. He maintains that the chief shortcoming of all of them is a failure to appreciate adequately the rich and complex structures of scientific theories. Appealing to the semantic conception of theories, he develops in considerable detail his own nondeductivist account of explanation. It is nondeductivist in two respects. In the first place, as remarked above, Suppe adopts the narrow construal of deductivism, according to which deductivism includes a commitment to the standard conception of theories as partially interpreted axiomatic systems. He rejects the standard conception in favor of the semantic conception, but this does not disqualify him as a deductivist in the broader sense. In the second place, however, he admits that scientific explanations of particular occurrences are possible in indeterministic contexts. This fact does disqualify him as a deductivist in the broader sense, and on this point I heartily agree. The accounts he offers of explanation *why,* explanation *how could,* and explanation *how did,* merit careful consideration.

How good is the case against explanatory deductive chauvinism advanced in the essays in this collection? Fairly strong, I think, but not nearly as conclusive as the case against *inferential* deductive chauvinism offered in this volume by Hempel, Kyburg, and me.

Conclusion

Quite possibly the best-known opposition to deductivism has come from philosophers inspired by the work of Thomas Kuhn, who has expressed opposition to the notion that the logic of science can be rationally reconstructed in any formal way. The choice among scientific theories, he claims, depends upon more than experimental evidence and logic. Among other things, the personal judgments of individual scientists are involved. This kind of critical approach to deductivism is not treated at length in the present volume, although Hempel gives a distinct nod in that direction at the conclusion of his paper.

The philosophers whose papers appear in this volume are all committed to the view that formal considerations are extremely fruitful in understanding the structure of scientific knowledge. Although they are not (with the possible exception of Giere) attempting to defend deductivism, they do not believe that rejecting deductivism means relinquishing all the tools that have achieved striking success in the hands of deductivists.[25]

Notes

1. A clear neo-Popperian account can be found in John Watkins, *Science and Scepticism* (Princeton, N.J.: Princeton University Press, 1984).
2. Wolfgang Stegmüller, *Probleme und Resultate der Wissenschaftstheorie und Analytischen Philosophie*, Band 4, Studienausgabe Teil E (Berlin and New York: Springer-Verlag, 1973).
3. G. H. von Wright, *Explanation and Understanding* (Ithaca, N.Y.: Cornell University Press, 1971).
4. In a fully formalized system—as studied by modern logic—none of the terms, not even those of logic and elementary arithmetic, have any meaning. Every term that occurs in such a system is either a primitive term or it is defined in terms of the primitive terms. No meanings are imported from outside. For our purposes, since we are concerned with the empirical sciences, it will not be necessary to deal with such fully formalized systems. The sort of axiomatic system I am describing might be called a *quasi-formal system*.
5. This is a semantic characterization of the concept of *logical consequence;* it is the concept that is pertinent to the uses of formal systems in empirical contexts. There is, as will be mentioned below, a purely syntactical concept of *logical consequence.* According to this concept, which does not involve truth in any way, one formula (a theorem) is a logical consequence of some other formulas (the axioms), if the former can be derived from the latter by formal manipulations permitted by the transformation rules of the system. The syntactical concept of logical consequence is of interest chiefly in dealing with fully formalized uninterpreted axiomatic systems.
6. For a clear exposition of the standard view of theories and a survey of the problems connected with it, see Carl G. Hempel, "The Theoretician's Dilemma," in his *Aspects of Scientific Explanation and Other Essays in the Philosophy of Science* (New York: The Free Press, 1965), pp. 173–226.
7. Both Giere and Suppe give hints that this is the thesis they reject— Giere in contrasting simple universal generalizations with mathematical equations; Suppe in referring to the rich structure of theories in comparison with the simple structure of laws.
8. See Carl B. Boyer, *The History of the Calculus and Its Conceptual Development* (New York: Dover, 1959), chap. 7.
9. It is important to keep in mind that, for all practical purposes, the syntactic concepts of deductive consequence and logical equivalence coincide, respectively, with the corresponding semantic concepts.
10. This point is taken from an unpublished letter from Coffa to Grünbaum; it can also be seen in Coffa's article "Probabilities: Reasonable or True?" *Philosophy of Science* 44 (1977):186–198.
11. See Wesley C. Salmon, *Scientific Explanation and the Causal Struc-*

ture of the World (Princeton, N.J.: Princeton University Press, 1984), chaps. 1 and 4, esp. pp. 84–97.

12. To be more accurate, we should refer to facts that can be described in terms of the antecedently understood vocabulary of Hempel's account, but for ease of introductory exposition I shall refer simply to observable facts.

13. See Carl G. Hempel, "Provisos," chap. 1 in this volume.

14. See Ronald N. Giere, "Laws, Theories, and Generalizations," chap. 2 in this volume.

15. Many science museums have Foucault pendulums to demonstrate the earth's rotation. A suitably pulsed electromagnet is sometimes used to keep the pendulum swinging for indefinite periods of time. A pendulum of this sort qualifies as a magnetically augmented pendulum—which shows that this concept is not merely a figment of a philosopher's imagination.

16. As Popper has frequently pointed out, we may be mistaken; observation does not yield infallible results.

17. This conclusion is also fallible. We cannot be absolutely certain either that we have correctly characterized the initial conditions or that our auxiliary assumptions are true.

18. I deny—as would Popper, I presume—that we have precognitive experience of future facts.

19. See Wesley C. Salmon, "Rational Prediction," chap. 3 of this volume.

20. In a series of four articles published in 1976, Grünbaum provides an extended and detailed critical *tour de force* against Popperian deductivism: "Is Falsifiability the Touchstone of Scientific Rationality? Karl Popper versus Inductivism," in R. S. Cohen, P. K. Feyerabend, and Marx Wartofsky, eds., *Essays in Memory of Imre Lakatos* (Dordrecht: Reidel, 1976), pp. 213–252; "Can a Theory Answer More Questions than One of Its Rivals?" *British Journal for the Philosophy of Science* 27 (1976):1–23; "Is the Method of Bold Conjectures and Attempted Refutations *Justifiably* the Method of Science?" *British Journal for the Philosophy of Science* 27 (1976):105–136; "*Ad Hoc* Auxiliary Hypotheses and Falsificationism," *British Journal for the Philosophy of Science* 27 (1976):329–362.

21. Carl G. Hempel and Paul Oppenheim, "Studies in the Logic of Explanation," *Philosophy of Science* 15 (1948):135–175. Reprinted, with a "Postscript," in Hempel, *Aspects of Scientific Explanation*, pp. 245–295.

22. J. Alberto Coffa, "Hempel's Ambiguity," *Synthese* 28 (1974):141–164, and "Probabilities: Reasonable or True?"

23. Michael Friedman, "Explanation and Scientific Understanding," *Journal of Philosophy* 71 (1974):5–19. See Philip Kitcher, "Explanation, Conjunction, and Unification," *Journal of Philosophy* 73 (1976):207–212, for a penetrating critique of Friedman's approach.

24. In speaking of *the cause* in this context, I am distinguishing between

a particular event that would normally be singled out as the cause and a host of background conditions that are required for the cause to be operative.

25. For my own part, I maintain that many of Kuhn's insights about the place of individual judgment in science can be incorporated into a reasonably formal explication of scientific confirmation if Bayes's theorem is brought into the picture. This point was briefly sketched in Wesley C. Salmon, "Carl G. Hempel on the Rationality of Science," *Journal of Philosophy* 80 (1983):555–562. The mathematical calculus of probability has, in my opinion, powerful resources to aid us in understanding scientific knowledge. This point is discussed at some length in Chap. VII of my book *The Foundations of Scientific Inference* (Pittsburgh, Pa.: University of Pittsburgh Press, 1967).

1. Provisos: A Problem Concerning the Inferential Function of Scientific Theories

Carl G. Hempel

The principal goal and the proudest achievement of scientific inquiry is the construction of comprehensive theories that give us an understanding of large classes of empirical phenomena and enable us to predict, to retrodict, and to explain them. These various functions of theories are usually regarded as having the character of inferences that lead, by way of theoretical principles, from sentences expressing initial and boundary conditions to statements describing the occurrences to be predicted, retrodicted, or explained.

In this paper*, I propose to examine a basic difficulty that faces this inferential construal of scientific theorizing and that has implications for some central issues in the philosophy of science. I will first present the problem by reference to a purely deductivist conception of theoretical reasoning and will then broaden its scope.

The Standard Deductivist Model

The best-known precise elaboration of a deductivist conception is provided by the so-called standard empiricist construal of theories and their application. It views a theory T as characterizable by an ordered pair consisting of a set C containing the basic principles of the theory and a set I of interpretative statements. Thus:

$$(1) \qquad\qquad T = <C,I>$$

The sentences, or formulas, of C serve to characterize the specific entities and processes posited by the theory (e.g., elementary particles and their interactions) and to state the basic laws to which they are assumed to conform. These sentences are formulated with the help of a theoretical vocabulary V_C whose terms refer to the kinds and characteristics of the theoretical entities and processes in question.

The sentences of the interpretative set I serve to link the theoretical scenario represented by C to the empirical phenomena to which the theory is to be applied. These phenomena are taken to be formulated in a vocabulary V_A that is antecedently understood, that is, available and understood independently of the theory. Thus, the sentences of I are said to provide partial interpretations, though not necessarily full definitions, of the theoretical terms in V_C by means of the antecedently understood terms of V_A. So-called operational definitions and reduction sentences in Carnap's sense may be viewed as special kinds of interpretative sentences.

By way of a simple example, assume that T is an elementary theory of magnetism whose theoretical vocabulary V_C contains such terms as "magnet," "north pole," "south pole," and whose theoretical principles include the laws of magnetic attraction and repulsion and the law that the parts of a magnet are magnets again, while the class I includes some operational criteria for the terms of V_C.

Consider now the following application of the theory. From the sentence "b is a metal bar to which iron filings are clinging" (S_A^1), by means of a suitable operational criterion contained in the set I, infer "b is a magnet" (S_C^1). Then, by way of theoretical principles in C, infer "If b is broken into two bars b_1 and b_2, then both are magnets and their poles will attract or repel each other" (S_C^2). Finally, using further operational criteria from I, derive the sentence "If b is broken into two shorter bars and these are suspended, by long thin threads, close to each other at the same distance from the ground, they will orient themselves so as to fall into a straight line" (S_A^2). (Note that V_A is here taken to contain not only predicates like "metal bar" but also individual constants such as "b.")

The basic structure thus attributed to a theoretical inference is suggested by the following schema, in which the notation $P \xrightarrow{Q} R$ is to indicate that R can be inferred from P by using sentences from Q as additional premises.

(2)

Thus, if the inferential steps in question are indeed all deductive, then the theory provides a deductive inference bridge leading from one V_A sentence, through the theoretical realm of C, to another V_A sentence. More precisely: S_A^1 in combination with the theory T deductively implies S_A^2; this in turn, is tantamount to saying that T deductively implies a corresponding V_A sentence, namely, the conditional $S_A^1 \supset S_A^2$.

Carnap and other logical empiricists assumed that the vocabulary V_A, which serves to describe the phenomena to be explained by the theory, consists of terms that are "observational" at least in a broad sense, that is, that they refer to features of the world whose presence or absence can be established by means of more or less direct observation. In recognition of the difficulties that face the notion of observability, I want to avoid any such assumption here. Indeed, I want to provide specifically for cases in which, as often happens, the vocabulary V_A was originally introduced in the context of an earlier theory. All that the standard construal needs to assume is that the phenomena for which the theory is to account are described by means of a vocabulary V_A that is "antecedently available" in the sense that it is well understood and is used with high intersubjective agreement by the scientists in the field. The interpretative sentences in I may then be viewed as interpreting the new terms introduced by the theory, that is, those in V_C, by means of the antecedently understood terms in V_A.

This deductivist construal[1] faces two basic difficulties. I will call them the problem of theoretical ascent and the problem of provisos. Let me spell them out in turn.

Inductive or Theoretical Ascent

The first inferential step in the schematic argument about the bar magnet presupposes that with the help of interpretative sentences belonging to part I of the theory of magnetism, S_C^1 is deducible from S_A^1. Actually, however, the theory of magnetism surely contains no general principle to the effect that when iron filings cling to a metal bar, then the bar is a magnet. The theory does not preclude the possibility, for example, that the bar is made of lead and is covered with an adhesive to which the filings stick, or that the filings are held in place by a magnet hidden under a wooden board supporting the lead bar. Thus, the theory does not warrant a deductive transition from S_A^1 to S_C^1. It is more plausible to assume that the theory contains an interpretative principle that is the converse of the one just considered, namely, that if a bar is a magnet, then iron filings will cling to it. But even this is not strictly correct, as will be argued shortly.

Hence, the transition from S_A^1 to S_C^1 is not deductive even if the entire theory of magnetism is used as an additional premise. Rather, the transition involves what I will call *inductive* or *theoretical ascent*, that is, a transition from a data sentence expressed in V_A to a theoretical hypothesis S_C^1 that would explain, by way of the theory of magnetism, what the data sentence describes.

This illustrates one of the two problems mentioned before that face a strictly deductivist construal of the systematic connections that a theory establishes between V_A-sentences, that is, between sentences describing empirical phenomena in terms of V_A. This problem has been widely discussed and various efforts have been made to resolve it by constructing theories of inductive reasoning that would govern such theoretical ascent. I will not consider those efforts here but will rather turn to the problem of provisos, which has not, it seems to me, been investigated in the same detail.

Provisos

Consider the third step in our example, the transition from S_C^2 to S_A^2. Again, the theory of magnetism does not provide interpretative hypotheses that would turn this into a strictly deductive inference. The theory clearly allows for the possibility that two bar magnets, suspended close to each other at the same level by fine threads, will not arrange themselves in a straight line. For example, if a strong magnetic field of suitable direction should be present in addition, then the bars would orient themselves so as to be parallel to each other; similarly, a strong air current would foil the prediction, and so forth.

The theory of magnetism does not guarantee the absence of such disturbing factors. Hence, the inference from S_C^2 to S_A^2 presupposes the additional assumption that the suspended pieces are subject to no disturbing influence, or, to put it positively, that their rotational motions are subject only to the magnetic forces they exert upon each other. (Incidentally, the explanatory inference mentioned a moment ago, from S_C^1 to S_A^1, presupposes an analogous tacit premise and thus is not deductive.) I will use the term "provisos" to refer to *assumptions* of the kind just illustrated, *which are essential, but generally unstated, presuppositions of theoretical inferences.*

Provisos are presupposed also in ostensibly deductive inferences that lead from one V_C-sentence to another. This holds, for example, in the inference from S_C^1 to S_C^2 in the case of the magnet: for if the breaking of the magnet takes place at a high temperature, the pieces may become demagnetized.

As a second example consider the application of the Newtonian theory of gravitation and of motion to a system of physical bodies like our solar system. In predicting, from a specification of the state of the system at a time t_0, subsequent changes of state, the basic idea is that the force acting on any one of the bodies is the vector sum of the gravitational forces exerted on it by the other bodies in accordance with the law of gravitation. That force then determines, via the second law of motion ($f = ma$), the resulting change of velocity and of position for the given body. But the quantity f in the second law is understood to be the *total* force acting on the given body, and the envisaged application of the theory therefore presupposes a proviso to the effect that the constituent bodies of the system are subject to no forces other than their mutual gravitational attraction. This proviso precludes not only gravitational forces that might be exerted by bodies outside the system but also any electric, magnetic, frictional, or other forces to which the bodies in the system might be subject. The absence of such forces is not, of course, vouchsafed by the principles of Newton's theory, and it is for this reason that the proviso is needed.

Escape by Interpretative Sentences of Probabilistic Form?

The foregoing considerations show in particular that when a theory contains interpretative sentences in the form of explicit definitions or of Carnapian reduction chains based on the antecedent vocabulary, the applicability of these sentences is usually subject to the fulfillment of provisos; they cannot be regarded as unequivocal complete or partial criteria of applicability for theoretical expressions.

This thought might suggest a construal of the interpretative sentences of a theory as expressing only probabilistic rather than strictly general connections between theoretically described states or events and certain associated manifestations or indicator phenomena described in antecedently available terms. Such a construal might seem to come closer to scientific usage and at the same time to obviate the need for provisos: for with probabilistic interpretation sentences, a theory would establish at best probabilistic connections between V_A sentences. And what would otherwise appear as occasional violations of provisos would be automatically anticipated by the merely probabilistic character of the theoretical inferences.

Interpretative sentences of probabilistic form have in fact been envisaged by several writers. Carnap did so in his essay "The Methodological Character of Theoretical Concepts" (1956), which is, I think, his earliest full presentation of the standard empiricist construal of theories. He argues there that many terms functioning in scientific theories cannot be regarded as linked to antecedent terms ("observational terms") by interpretative sentences ("rules of correspondence") of strictly universal form. For such sentences would specify strictly necessary or sufficient observational conditions of applicability for the theoretical terms, whereas scientists, Carnap argues, will treat such conditions not as strictly binding but as qualified by an "escape clause" to the effect that the observational criteria hold "unless there are disturbing factors" or "provided the environment is in a normal state" (p. 69). Such escape clauses clearly have the character of provisos in the sense adumbrated above. Carnap views them as probabilistic qualifiers functioning in interpretative sentences for theoretical terms. These sentences would state probabilistic rather than strictly necessary or sufficient connections between theoretical formulas and V_A- sentences. Indeed, while Carnap countenances dispositional terms, linked to V_A by strict reduction chains, he suggests that the terms characteristic of scientific theories have only probabilistic links to the observational basis (pp. 49, 72).

But while Carnap thus explicitly eschews a purely deductivist construal of the inferential function of theories, he does not specify the form of the probabilistic interpretation sentences he envisages. Indeed, in response to a proposal by Arthur Pap (1963, Sec. II) concerning probabilistic reduction sentences, Carnap remarks: "it seems to me that for the time being the problem of the best form for [interpretative sentences] has not yet been sufficiently clarified" (1963, p. 950).

However that may be, a probabilistic construal of provisos faces the difficulty that scientific theories do not, in general, provide probabilistic laws that would obviate the need for provisos. Consider, for example, the interpretative sentences that would be required for the term "magnet." They would have to take the form "In cases in which iron filings stick to a metal bar, the probability of the bar's being a magnet is p_1," or, for inferences in the opposite direction, "Given that a metal bar is magnetic, the probability that iron filings will cling to it is p_2." But surely, the theory of magnetism contains no sentences of this kind; it is a matter quite beyond the theory's scope to state how frequently air currents, further magnetic fields, or other factors will interfere with the effect in question. It seems to me that no scientific theory provides probabilistic interpretation statements of this sort, whose application is not itself subject to provisos.

The same basic consideration applies also, I think, where no well-developed and sharply formulated theories are available. For example, a probabilistic construal cannot avoid the need for provisos in the application of theoretical sentences linking psychological states or events to their behavioral manifestations.

Some Consequences of the Need for Provisos

The conclusion that a scientific theory even of nonprobabilistic form does not, in general, establish deductive bridges between V_A-sentences has significant consequences for other issues in the philosophy of science. I will briefly indicate four of these: the idea of falsifiability, the significance of so-called elimination programs for theoretical terms, the instrumentalist construal of scientific theories, and the idea of "the empirical content" of a theory.

Falsifiability

One obvious consequence of the need for provisos is that normally a theory is not falsifiable by V_A-sentences; otherwise, it would deductively imply the negations of the falsifying V_A-sentences, which again are V_A-sentences. This consideration differs from the

Duhem-Quine argument, which is that individual hypotheses cannot be falsified by experiential findings because the deduction from the hypothesis of falsifying V_A sentences requires an extensive system of background hypotheses as additional premises, so that typically only a comprehensive set of hypotheses will entail or contradict V_A sentences. The argument from provisos leads rather to the stronger conclusion that even a comprehensive system of hypotheses or theoretical principles will not entail any V_A sentences because the requisite deduction is subject to provisos.

Note that a proviso as here understood is not a clause that can be attached to a theory as a whole and vouchsafe its deductive potency by asserting that in all particular situations to which the theory is applied, disturbing factors are absent. Rather, a proviso has to be conceived as a clause that pertains to some particular application of a given theory and asserts that in the case at hand, no effective factors are present other than those explicitly taken into account.

Elimination Programs for Theoretical Terms

The need for provisos also has a bearing on the so-called elimination programs for theoretical terms. These programs are of particular significance for philosophical qualms about the use, in scientific theories, of terms that are not explicitly defined by means of an antecedently understood vocabulary.

The ingenious and logically impeccable methods designed by Frank P. Ramsey and by William Craig[2] circumvent these qualms by showing that the use of theoretical expressions can always be avoided in the following sense: If a theory T consisting of two sentence classes C and I (as characterized above) does yield deductive connections between certain V_A sentences, then it is possible to formulate a corresponding theory (class of sentences) T_A such that

(i) T_A is expressed in terms of V_A alone;

(ii) T_A is logically implied by T;

(iii) T_A entails "$S_A^1 \supset S_A^2$" (and in this sense establishes a deductive bridge from S_A^1 to S_A^2) if and only if T entails "$S_A^1 \supset S_A^2$."[3]

If the function of a theory is taken to consist in establishing deductive bridges among V_A sentences, then the theory T_A, which avoids the use of theoretical terms, might be called functionally equivalent to the theory T. This result might suggest the reassuring conclusion that, in principle, the use of theoretical expressions can always be avoided without any change in the "empirical content" of a theory as it is expressed by the class of V_A sentences deducible from it, and that talk in terms of theoretical expressions is just a convenient *façon de parler* about matters that are fully expressible in the antecedently understood vocabulary V_A. Analogously, it may seem that all the problems about theoretical ascent and provisos simply disappear if T is replaced by its functional equivalent T_A.

This impression is illusory, however. For a theory T_A constructed from T in the manner of Ramsey or of Craig yields deductive connections between V_A sentences if and only if T yields such connections: and scientific theories do not, in general, satisfy this condition. The need for provisos precludes the general avoidability of theoretical expressions by those elimination methods.

The verdict does not hold, however, if the provisos qualifying the inferential applications of a theory are themselves expressible in the antecedent vocabulary. For if P_A is such a proviso governing the transition, by means of T, from S_A^1 to S_A^2, then T entails the sentence $(P_A \cdot S_A^1) \supset S_A^2$ and thus establishes a deductive bridge between two V_A sentences.

But it seems that, in general, the requisite provisos cannot be expressed in terms of V_A alone. In the case of the theory of magnetism referred to earlier, the provisos may assert, for example, the absence of other magnetic fields or of disturbing forces, etc., and will then require at least the use of terms from V_C in their formulation.

Provisos and the Instrumentalist Construal

The preceding considerations analogously cast some doubts on the instrumentalist conception of theories as purely inferential devices that, from an input in the form of V_A sentences, generate an output of other V_A sentences. For the need for provisos shows that theories do not render this service. In each particular case, the applicability of the theoretical instrument would be subject to the condition that the pertinent provisos are fulfilled; and the assertion that they are fulfilled could not just be added to the input into the theoretical calculating machine, for that assertion would not generally be expressible in V_A.

Thus, if a theory is to be thought of as a calculating instrument that generates new V_A sentences from given ones, then it must be conceived as supplemented by an instruction manual specifying that the instrument should be used only in cases in which certain provisos are satisfied. But the formulation of those provisos will make use of V_C and perhaps even of terms not contained in V_C. Thus, one has to check whether certain empirical conditions not expressible in V_A are satisfied: and that surely provides a tug away from instrumentalism and in the direction of realism concerning theoretical entities.

Provisos and "The Empirical Content" of a Theory

Similar questions arise in regard to the notion of the experiential "cash value" or "empirical content" of a theory as represented by the set of all V_A-sentences entailed by the theory. Note first, and incidentally, that thus construed the empirical content of a theory is relative to the vocabulary V_A that counts as antecedently available, so that one would properly have to speak, not of "the" empirical content of T, but of the V_A-content of T. But the point here to be made is rather that usually a theory does not entail V_A-sentences and the proposed construal of empirical content fails.

To be sure, there are some deductive theoretical inferences that presuppose no provisos; for example, the inference, mediated by the law of gravitation, from a sentence S^1 specifying the masses and the distance of two bodies to a sentence S^2 specifying the gravitational attraction that the bodies exert upon each other. But the further theoretical inference from S^2 to a sentence S^3 specifying the accelerations the bodies will undergo requires a proviso to the effect that no other forces act upon the bodies. If S^2 and S^3 are represented as theoretical sentences, then we have here an example of the need for provisos not only in establishing theoretical inference bridges between V_A-sentences and V_C-sentences but also in building such bridges between sentences expressed solely in terms of V_C. We will shortly return to this point.

Further Thoughts on the Character
of Provisos

How might the notion of a proviso be further illuminated? To say that provisos are just *ceteris paribus* clauses is unhelpful, for the idea of a *ceteris paribus* clause is itself vague and elusive. "Other things being equal, such and such is the case." What other things, and equal to what? How is the clause to function in theoretical reasoning? Provisos do not call for the equality of certain things, but for the absence of disturbing factors.

Provisos might rather be viewed as *assumptions of completeness*. The proviso required for a theoretical inference from one sentence, S^1, to another, S^2, asserts, broadly speaking, that in a given case (e.g., in that of the metal bar considered earlier) no factors other than those specified in S^1 are present that could affect the event described by S^2.

For example, in the application of Newtonian theory to a double star it is presupposed that the components of the system are subject to no forces other than their mutual gravitational attraction and, hence, that the specification given in S^1 of the initial and boundary conditions that determine that gravitational attraction is a complete or exhaustive specification of all the forces affecting the components of the system.

Such completeness is of a special kind. It differs sharply, for example, from that invoked in the requirement of complete or total evidence. This is an epistemological condition to the effect that in a probabilistic inference concerning, say, a future occurrence, the total body of evidence available at the time must be chosen as the evidential basis for the inference.[4]

A proviso, on the other hand, calls not for epistemic but for ontic completeness: the specifics expressed by S^1 must include not all the information available at the time (information that may well include false items) but rather all the factors present in the given case that in fact affect the outcome to be predicted by the theoretical inference. The factors in question might be said to be those that are "nomically relevant" to the outcome, that is, those on which the outcome depends in virtue of nomic connections.

Consider once again the use of Newtonian theory to deduce, from a specification S^1 of the state of a binary star system at time t_1, a specification S^2 of its state at t_2. Let us suppose, for simplicity, that S^1 and S^2 are couched in the language of the theory; this enables us to disregard the problem of the inductive ascent from astronomical observation data to a theoretical redescription in terms of masses, positions, and velocities of the two objects.

The theoretical inference might then be schematized thus:

(3) $(P \cdot S^1 \cdot T) \rightarrow S^2$

where P is a proviso to the effect that apart from the circumstances specified in S^1, the two bodies are, between t_1 and t_2, subject to no influences, from within or from outside the system, that would affect their motions. The proviso must thus imply the absence, in the case at hand, of electric, magnetic, and frictional forces; of radiation pressure; and of any telekinetic, angelic, or diabolic influences.

One may well wonder whether this proviso can be expressed in the language of celestial mechanics at all, or even in the combined languages of mechanics and other physical theories. At any rate, neither singly nor jointly do those theories assert that forces of the kinds they deal with are the only kinds by which the motion of a physical body can be affected. A scientific theory propounds an account of certain kinds of empirical phenomena, but it does not pronounce on what other kinds there are. The theory of gravitation neither asserts nor denies the existence of nongravitational forces, and it offers no means of characterizing or distinguishing them.

It might seem, therefore, that the formulation of the proviso transcends the conceptual resources of the theory whose deductive applicability it is to secure. That, however, is not the case in the example at hand. For in Newton's second law, $f = ma$, "f" stands for the *total* force impressed on the body; and our proviso can therefore be expressed by asserting that the total force acting on each of the two bodies equals the gravitational force exerted upon it by the other body; and the latter force is determined by the law of gravitation.[5] But the application of the theory to particular cases is clearly subject again to provisos to the effect that in computing the total force, all relevant influences affecting the bodies concerned have been taken into account.

When the application of a theory to empirical subject matter is schematically represented in the form (3) with the proviso P as one of the premises, it must be borne in mind that the language and the specific form in which P is expressed are left quite vague. The notation is meant to be not a sharp explication but rather a convenient way of referring to the subject at issue in the context of an attempt to shed some further light on it.

Note that the proviso P does not include clauses to the affect that the establishment of S^1 has not been affected by errors of observation or measurement, by deceit or the like: that is already implied by the premise S^1 itself, which trivially asserts that S^1 is true. The proviso is to the effect not that S^1 is true but that it states the *whole* truth about the relevant circumstances present.

Note further that the perplexities of the reliance on provisos cannot be avoided by adopting a structuralist, or nonstatement, conception of theories broadly in the manner of J. D. Sneed (1979) and Wolfgang Stegmüller (1976, esp. chap. 7). That conception construes theories not as classes of statements but as deductively organized systems of statement functions, which make no assertions and have no truth values. But such systems are presented as having empirical models; for example, the solar system might be claimed to be a model of a structuralist formalization of Newtonian celestial mechanics. But a formulation of this claim, and its inferential application to particular astronomical occurrences, again clearly assumes the fulfillment of pertinent provisos.

Methodological Aspects of Provisos

The elusive character of proviso clauses raises the question how a theoretical inference of type (3) can be applied to particular occurrences, and more specifically, on what grounds proviso P may be taken to be satisfied or violated in specific cases.

There are circumstances that provide such grounds. If the theory T has strong previous support, but its application to a new case yields incorrect predictions S^2, then doubts may arise about S^1; but in the absence of specific grounds for such doubts, a violation of P—that is, the presence of disturbing factors—may suggest itself. If this conjecture can be expressed in the language of the theory T, and if replacing S^1 by a correspondingly modified sentence $S^{1'}$ yields successful predictions, then this success will constitute grounds for attributing the predictive failure of the original theoretical inference to a violation of its proviso clause.

Thus, the failure of Newton's otherwise highly successful theory to predict certain perturbations in the orbit of Uranus in terms of the gravitational attraction exerted on it by the sun and by the planets known before 1846 led to the conjecture of a proviso violation, namely, that Uranus was subject to the additional attraction of a hitherto unknown planet—a conjecture borne out by the subsequent discovery of Neptune.

Sometimes predictive failure of a theory is attributed to proviso violations even though the presumably disturbing factors cannot be adequately specified. Consider, for example, the controversy between Robert A. Millikan and Felix Ehrenhaft over the results of the famous experiments in which Millikan measured the rates at which small, electrically charged oil drops rose and sank in an electric field between two horizontal, electrically charged metal plates. From those rates he computed, by means of accepted theoretical principles, the size of the charges of those oil drops and found that all of them were integral multiples cf a certain minimum charge e, whose numerical value he specified. Millikan presented his findings as evidence for the claim that electricity had an atomistic structure and that the atoms of electricity all had the specified charge e.

Ehrenhaft objected that in similar experiments, he had found individual charges that were not integral multiples of Millikan's value e and that were, in fact, often considerably smaller than e, suggesting the existence of "sub-electrons."[6] Ehrenhaft accordingly rejected Millikan's theoretical claims T on grounds of predictive failure.

Millikan replied in careful detail. Referring to difficulties he had encountered in his own work, he argued that Ehrenhaft's deviant results could be due to disturbing factors of various kinds. Among them, he mentioned the possibility that tiny dust particles had settled on the falling oil droplets, thus changing the total force acting on them; the possibility that evaporation had reduced the mass of an observed drop; the possibility that the strength of the electric field had decreased as a result of battery fatigue, and so forth.

Ehrenhaft repeated his experiments, taking great pains to screen out such disturbing factors, but he continued to obtain deviant findings. The sources of these deviations have never been fully determined; in fact, Ehrenhaft's results turned out not to be generally reproducible. Millikan's ideas, on the other hand, were sustained in various quite different applications. Thus, Ehrenhaft's claims were gradually disregarded by investigators in the field, and Millikan won the day and the Nobel prize. Interestingly, as has been pointed out by Gerald Holton (1978, esp. pp. 58–83), Millikan himself had recorded in his laboratory diaries several sets of quite deviant measurements, but he had not published them, attributing them to disturbing factors of various kinds and sometimes not even offering a guess as to the nature of those factors.

But evidently, it cannot be made a *general* policy of scientific research to attribute predictive failures of theoretical inferences to the violation of some unspecified proviso; for this "conventionalist strategem," as Karl Popper has called it, would deprive a theory of any predictive or explanatory force.[7]

I think that at least in periods of what Thomas Kuhn calls normal science, a search for disturbing influences will consider only factors of such kinds as are countenanced by one or another of the currently accepted scientific theories as being nomically relevant to the phenomena under consideration. Thus, if a prediction based on Newtonian mechanics fails, one might look for disturbing gravitational, electric, magnetic, and frictional forces and for still some other kinds but not for telekinetic or diabolic ones. Indeed, since there are no currently accepted theories for such forces, we would be unable to tell under what conditions and in what manner they act; consequently, there is no way of checking on their presence or absence in any particular case.

The mode of procedure just mentioned is clearly followed also in experiments that require the screening out of disturbing outside influences—for example, in experimental studies of the frequency with which a certain kind of subatomic event occurs under specified conditions. What outside influences—such as cosmic rays—would affect the frequency in question, and what shielding devices can serve to block them and thus to ensure satisfaction of the relevant proviso, is usually determined in the light of available scientific knowledge, which again would indicate no way of screening out, say, telekinetic influences.

If a theory fails to yield correct predictions for a repeatable phenomenon by reference to factors it qualifies as relevant, then certain changes within that theory may be tried, changes introducing a new kind of nomically relevant factor. Wilhelm Röntgen's discovery of a photographic plate that had been blackened while lying in a closed desk drawer is, I think, a case in point; it led to the acknowledgment of a new kind of radiation.

Finally, persistent serious failures of a theory may lead to a revolution in Kuhn's sense, which places the phenomena into a novel theoretical framework rather than modifying the old one by piecemeal changes. In this case, the failures of the earlier theory are not attributed to proviso violations; indeed it is quite unclear what such an attribution would amount to.

Consider a theoretical inference that might have been offered some 250 years ago on the basis of the caloric fluid theory of heat or the phlogiston theory of combustion. The relevant provisos would then have to assert, for example, that apart from the factors explicitly taken into account in the inference, no other factors are present that affect, say, the flow of caloric fluid between bodies or the degree of dephlogistication of a body. But from our present vantage point, we have to say that there are no such substances as caloric fluid or phlogiston, and that therefore there could be no proper proviso claim of the requisite sort at all. Yet it appears that the claims and the inferential applications of any theory have to be understood as subject to those elusive provisos.

There is a distinct affinity, I think, between the perplexing questions concerning the appraisal of provisos in the application of scientific theories and the recently much-discussed problems of theory choice in science. As Kuhn in particular has argued in detail, the choice between competing theories is influenced by considerations concerning the strength and the relative importance of various desirable features exhibited by the rival theories; but these considerations resist adequate expression in the form of precise explicit criteria. The choice between theories in the light of those considerations, which are broadly shared within the scientific community, is not subject to, nor learned by means of, unambiguous rules. Scientists acquire the ability to make such choices in the course of their professional training and careers, somewhat in the manner in which we acquire the use of our language largely without benefit of explicit rules, by interaction with competent speakers.

In the context of theory choice, the relevant idea of the superiority of one theory to another has no precise explication, yet it is not arbitrary, for its use is strongly affected by considerations shared by scientific investigators. Likewise, in the inferential application of theories to empirical contexts, the idea of the relevant provisos has no precise explication, yet it is by no means arbitrary, for its use appears to be significantly affected by considerations akin to those affecting theory choice.

Notes

*This paper is based upon work supported by National Science Foundation Grant No. SES 80-25399.

1. I have limited myself here to a schematic account of those features of the empiricist model which are of relevance to the problems to be discussed subsequently. For fuller expositions and critical discussions, and for references to the extensive literature, see, for example, Carnap 1956; Carnap 1966, chaps. 23–26; Feigl 1970; Hempel 1958, 1969, and 1970; Putnam 1962; and Suppe 1974, a comprehensive study that includes a large bibliography.

2. For details, see Ramsey 1931, Sec. IXA, "Theories"; Carnap 1966, chap. 26, "The Ramsey Sentence"; Craig 1956; Putnam 1965; and Hempel 1965, pp. 210–217.

3. The theory T_A obtainable by Ramsey's method is quite different, in other respects, from that generated by Craig's procedure. But the differences are irrelevant to the point under discussion here.

4. Cf. Carnap 1950, pp. 211–213, 494.

5. I am indebted to Michael Friedman for having pointed this out to me.

6. Millikan gives a detailed account of his investigations in Millikan 1917; he discusses Ehrenhaft's claims in chap. 8. The controversy is examined in a broader scientific and historical perspective in Holton 1978.

7. See, for example, Popper 1962, pp. 33–39, and Stegmüller 1976, chap. 14.

References

Carnap, Rudolf. 1950. *Logical Foundations of Probability.* Chicago: University of Chicago Press.

———. 1956. "The Methodological Character of Theoretical Concepts." In H. Feigl and M. Scriven, eds., *Minnesota Studies in the Philosophy of Science,* vol. 1, pp. 38–76. Minneapolis: University of Minnesota Press.

———. 1963. "Arthur Pap on Dispositions." In P. A. Schilpp, ed., *The Philosophy of Rudolf Carnap,* pp. 947–952. LaSalle, Ill.: Open Court.

———. 1966. *Philosophical Foundations of Physics.* New York and London: Basic Books.

Craig, William. 1956. "Replacement of Auxiliary Expressions," *Philosophical Review* 65:38–55.

Feigl, Herbert. 1970. "The 'Orthodox' View of Theories: Remarks in Defense as Well as Critique." In M. Radner and S. Winokur, eds., *Minnesota Studies in the Philosophy of Science,* vol. 4, pp. 3–16. Minneapolis: University of Minnesota Press.

Hempel, Carl G. 1958. "The Theoretician's Dilemma." In H. Feigl, M. Scriven, and G. Maxwell, eds., *Minnesota Studies in the Philosophy of Science,* vol. 2, pp. 37–98. Minneapolis: University of Minnesota Press. Reprinted in Carl G. Hempel, *Aspects of Scientific Explanation and*

Other Essays in the Philosophy of Science, pp. 173–226; New York: The Free Press, 1965.

————. 1969. "On the Structure of Scientific Theories." In Carl G. Hempel et al., *Isenberg Memorial Lecture Series 1965–1966,* pp. 11–38. East Lansing: Michigan State University Press.

————. 1970. "On the 'Standard Conception' of Scientific Theories." In M. Radner and S. Winokur, eds., *Minnesota Studies in the Philosophy of Science,* vol. 4, pp. 142–163. Minneapolis: University of Minnesota Press.

Holton, Gerald. 1978. "Subelectrons, Presuppositions, and the Millikan-Ehrenhaft Dispute." In Gerald Holton, *The Scientific Imagination,* pp. 25–83. Cambridge: Cambridge University Press, 1978.

Millikan, Robert A. 1917. *The Electron.* Chicago: University of Chicago Press. Facsimile edition, 1963.

Pap, Arthur. 1963. "Reduction Sentences and Disposition Concepts." In P. A. Schilpp, ed., *The Philosophy of Rudolf Carnap,* pp. 559–597. LaSalle, Ill.: Open Court.

Popper, Karl. 1962. *Conjectures and Refutations.* New York: Basic Books.

Putnam, Hilary. 1962. "What Theories Are Not." In E. Nagel, P. Suppes, and A. Tarski, eds., *Logic, Methodology, and Philosophy of Science,* pp. 240–251. Stanford, Calif.: Stanford University Press.

————. 1965. "Craig's Theorem," *The Journal of Philosophy* 62:251–260.

Ramsey, Frank P. 1931. *The Foundations of Mathematics and Other Logical Essays,* ed. R. B. Braithwaite. London: Routledge & Kegan Paul, 1931. Paperback reprint, Patterson, N.J.: Littlefield, Adams, 1960.

Sneed, J. D. 1979. *The Logical Structure of Mathematical Physics,* 2d ed. Dordrecht and Boston: Reidel.

Stegmüller, Wolfgang. 1976. *The Structure and Dynamics of Theories.* New York, Heidelberg, Berlin: Springer-Verlag.

Suppe, Frederick. 1974. "The Search for Philosophic Understanding of Scientific Theories." In Frederick Suppe, ed., *The Structure of Scientific Theories,* pp. 3–232. Urbana: University of Illinois Press. Second edition, with an "Afterword," 1977.

2. Laws, Theories, and Generalizations

Ronald N. Giere

The recent history of the philosophy of science is subject to conflicting interpretations. A common view is that logical empiricism reached its high-water mark around 1960, and during the succeeding generation was supplanted by a tradition stemming from the early works of Paul Feyerabend, Norwood Russell Hanson, Thomas Kuhn, Stephen Toulmin, and others. In this transition, analysis of the structure of science in terms of the logical structure of the content of science, particularly its laws and theories, supposedly gave way to discussion—which is still vigorous—of the historical development of concepts or "research traditions." Similarly, it is often claimed, analysis of the rationality of science in terms of (inductive) logical relations between evidence statements and hypotheses gave way to study of the rationality of larger historical units—"paradigms" or "research programmes."[1]

Although commonly held, the above account of recent developments in the philosophy of science ignores a large part of the literature of the past twenty years. It is true that a majority of philosophers of science would now reject most of the philosophical doctrines of logical empiricism—the verifiability theory of meaning, the theory/observation distinction, foundationism, etc. Yet a great many philosophers of science have continued to employ a variety of formal techniques to analyze both the content and the methods of various sciences.[2] Thus, while the existence or importance of larger historical processes is not necessarily denied, the idea that the analysis of science is to be done solely in terms of such processes is rejected.

I will resist the temptation to argue the general merits of a "new formalism" or a "new deductivism" in the philosophy of science. My purpose here shall be merely to illustrate how a continuity of concern with the formal structure of scientific hypotheses may be combined with a departure from central doctrines of logical empiricism. To do this I will focus on a problem that has concerned Carl G. Hempel (and several others) for at least a decade. I shall not claim that the problem cannot be "solved" within Hempel's modified logical empiricist framework. Rather, I shall claim that the problem is "better solved" within a different, though no less formal, philosophical framework.

Hempel's Provisos

Although he is rightfully regarded as the foremost exponent of latter-day logical empiricism, Hempel has also uncovered some of the most difficult problems for standard logical empiricist principles. Of course his aim has always been to strengthen the basic philosophical position, but such intellectual honesty may lead to problems that actually undermine one's own position. This may be the case with the problems raised by "provisos."[3]

Hempel typically discusses provisos in connection with the use of laws for prediction or explanation. But such contexts seem not to be essential to his discussion since in either case it is the truth of the law itself rather than its role in prediction or explanation that is called into question. It seems simpler, therefore, to focus directly on the status of laws themselves.

As an example let us take the familiar "law of the pendulum," which gives the period of oscillation (T) as a funtion of the length (l) and the strength of a uniform gravitational field (g).

$$(1) \qquad\qquad T \,=\, 2\pi \sqrt{l/g}$$

On the standard logical empiricist construal of scientific laws, (1) is not itself a law. A law is a universal generalization. To get the corresponding law let us define the predicate Tx to mean "having the property specified in (1)." Then let Px be the property of being a pendulum, where this is spelled out in "observational" talk about suspended weights free to swing through small arcs, etc. The law in question, then, is the statement:

(L) $(x)(Px \supset Tx)$.

Now what is the status of this "law"?

It has often been pointed out that statements like (L) typically are not strictly true. They assume various idealizations and approximations, for example, that there is no air friction and that the arc of swing is sufficiently small that sine(θ) is approximately equal to θ. But such facts have not generally been regarded as posing serious difficulties for the logical empiricist account of scientific laws. This account has itself been taken to be only a "first approximation" to be refined later. The problem Hempel raises is more serious. It is not just that (L) is not quite true but that it is grossly mistaken.

Imagine a pendulum consisting of an iron bob on a string. Directly under the rest position of the bob is a fairly strong magnet. If the pendulum is now set to swinging, its period may differ considerably from that given by (L), the exact amount of the discrepancy depending on the strength of the magnet. This problem cannot be passed over by saying that after all (L) is only "approximately" true. If we take (L) at face value, we shall have to say simply that it is false. Hempel's response has been that we not take (L) at face value. Rather, he says, laws are to be understood as containing implicit "provisos" that qualify them. Thus (L) is to be interpreted as containing the proviso that the bob is not magnetic or that there is no magnet underneath. Of course it must also contain many other implicit qualifications as well.

Hempel's view, then, is that laws do not in fact have the form of simple generalizations like (L) but of qualified generalizations better represented as

(QL) $(x)(Px \cdot Qx \supset Tx)$,

where Qx expresses the necessary provisos.

No one is more aware than Hempel of the immediate problems raised by this view. The problems he finds are both logical and epistemological. The epistemological problems all stem from the obvious fact that a qualified generalization is much harder to justify empirically than a simple, unqualified generalization. These are serious problems, but not as serious as the logical problems Hempel raises. I shall confine my discussion to the latter.

One immediate problem that Hempel faced was that qualified laws are incapable of being written down explicitly simply because the number of provisos implicit in any law is indefinitely large. There seems to be no obvious alternative to just accepting this incompleteness as an essential feature of scientific laws. A worse danger is that such a construal of laws renders them empirically vacuous. Unless there are some restrictions on the number and kind of qualifications allowed, any pendulum could satisfy the law. One need only construct a suitable "proviso." The problem is to formulate the needed restrictions without rendering the law completely trivial.[4] Finally, not only may some provisos be unknown; they may not even be expressible in the language of the relevant theory. This is in fact the case with the pendulum example. The relevant theory is classical mechanics, but the proviso refers to magnetic forces that were not explained theoretically until the nineteenth century. It is stretching the notion of implicit qualifications to say that they may not even be formulatable in terms of any then current theory.

It is impossible to prove that these problems cannot be satisfactorily resolved within the liberal logical empiricist framework Hempel has developed. Nevertheless—and here the Kuhnian categories seem very appropriate—it is possible to view these problems not merely as puzzles but as genuine anomalies. And there is no more effective way of realizing this possibility than to develop an alternative point of view in which the puzzles simply do not arise.

Theories, Generalizations, and Laws

There are a number of assumptions that form the philosophical background to Hempel's concerns about provisos. Among these assumptions are (i) that laws are empirical universal generalizations, and (ii) that theories are deductively related sets of empirical statements, at least some of which are universal generalizations. These assumptions are not peculiar to Hempel, nor to logical empiricism. They have been part of philosophical theories of science since Aristotle. Reinforced by the Euclidean tradition, they survived the seventeenth-century scientific revolution with renewed vigor. Logical empiricism is simply the most recent philosophy to give form to these Aristotelian ideals.

There is a minority tradition that regards theories as being more like definitions than sets of empirical statements. I expect that this minority viewpoint has a long history, but I am not prepared to document that history here. Patrick Suppes popularized this idea twenty-five years ago and his position has recently been further developed by Joseph D. Sneed and Wolfgang Stegmüller. Bas van Fraassen and Frederick Suppe have also developed versions of the general viewpoint, drawing inspiration from Everet Beth and John von Neumann respectively. I strongly prefer an intensionalist, state-space approach along the lines laid down by van Fraassen and Suppe, but these differences are not terribly relevant to the present discussion. More important are the similarities stemming from a model-theoretic rather than proof-theoretic approach to logic and mathematics.[5]

In presenting this alternative point of view, it is helpful to introduce the notion of a "theoretical model." A theoretical model is an abstract structure that may be characterized in a variety of ways utilizing a variety of linguistic resources. The use of theoretical models in the empirical sciences requires that one be able to identify elements of the abstract structure with elements of real empirical systems. Assuming the required identifications can be made, the characterization of a theoretical model becomes a definition of a kind of real system. Such definitions can be used to make empirical claims about particular real systems; namely, that they have the structure of the model in question. Although such claims may be internally very complex, they have the form of singular statements referring to individual real systems.

With the ability to make singular claims, generalizations come naturally. For any class of independently identifiable systems, it can be claimed that all members of that class exhibit the defined structure. Such classes can be of any size: small, large but finite, or potentially infinite. I will use the term "theoretical hypothesis" for any claim that attributes a specified theoretical structure to real systems. A theoretical hypothesis, then, can be of any scope, from a statement about a single system to a statement about a potential infinity of systems.

It is tempting to use the term "theory" to refer to characterizations of theoretical models. This temptation should be resisted, I think, just because the word "theory" carries such strong connotations of something that can be empirically true or false. In developing a theory of science one should not lightly go against well-established usage. However, conjunctions of singular theoretical hypotheses can be true or false, and these are likely candidates for the honorific title of "theory." We shall therefore say that theories are theoretical hypotheses of more or less restricted scope. This accords well with established usage. Newtonian mechanics, for example, is the theory that all systems of material bodies have the structure characterized by Newton's laws.[6]

What then are "laws"? Laws, on this account, are statements that characterize the structure of a theoretical model. Whenever the elements of a theoretical model are identified with physical properties, laws then describe constraints on the state or dynamic development of a physical system. The Boyle-Charles gas law, for example, is to be understood as asserting that the pressure, volume, and temperature of an ideal gas are constrained to satisfy the relationship $PV = nRT$. Note that this statement is not to be understood as a universal generalization concerning real gases, but as part of the characterization of what it is to be an ideal gas.

When referring to a theoretical model in formulating a theoretical hypothesis about an individual real system, the distinction between laws and initial conditions is fundamental. A statement of initial conditions merely describes the state of the system at some designated time. The laws say how that state is related to other possible states ("laws of coexistence") or to possible states at other times ("laws of succession"). This distinction thus helps to clarify the idea that laws describe the "structure" of a system and not merely its state, or even the whole succession of states that constitutes its total history.

It helps to fix these ideas if one considers examples from sciences other than physics. Biology provides a good touchstone since its status as a science has often been questioned. Today the failure of any model of science to accord full scientific status to biology would be grounds not for questioning the status of biology but for rejecting that model of science.

In the nineteenth and in the early twentieth century, biologists looked for the "laws of evolution" among existing species and in the fossil record. Bergmann's law, for example, states that for species of warm-blooded vertebrates, races living in cooler climates are larger than races living in warmer climates. Contrary to what some empiricist philosophers have claimed, the trouble with such laws is not that they fail to apply throughout the universe but that they hover between falsity and vacuity even when restricted to life on Earth.

If stated with sufficient precision to be informative, there are always some exceptions, for example, burrowing mammals. If stated with sufficient generality to avoid exceptions, they cease to be usefully informative. From the present perspective, the reason these "laws" have such a dubious status is that they are attempts to generalize over what are partly initial conditions (including the environment) of evolving systems of organisms. At best they express rough statistical generalizations reflecting the average selective forces of different environments on similar populations of organisms.

The real laws of evolution are the laws of the theoretical models used in formulating the hypotheses of evolutionary theory. These are, very briefly, (i) random variation in phenotypic traits, (ii) differential fitness as a function of phenotypic traits, and (iii) heritability of phenotypic traits. Any system of organisms satisfying these laws will evolve. The theory of evolution is the hypothesis that all (or at least most) species of organisms on Earth evolved, and are now evolving, by processes exhibiting a structure characterized by these laws.

Provisos Revisited

Now let us examine Hempel's problem from this new perspective. First off, we note that the law of the pendulum is better represented by equation (1) than by the formula (L). Indeed, we can construct a theoretical model that applies to pendulums. (This model is of course a special case of the models of classical mechanics.) In rough terms this model can be characterized as follows:

A system consisting of a suspended weight is a classical pendulum if and only if its period is given by equation (1).

It is then a theoretical hypothesis whether any particular suspended weight is indeed a classical pendulum.

The minor problem that no swinging weight exactly satisfies equation (1) has an easy solution. For any particular pendulum we can say that its behavior approximates that of a classical pendulum to within specified limits on its period of oscillation. This formulation makes it clear that approximation is not an absolute notion but one that presupposes a specification of the feature in question (e.g., the period) and a range of variation (e.g., one percent). It is thus a nontrivial generalization that many swinging weights are, to a high degree of approximation, classical pendulums.

The kind of problem that led Hempel to introduce provisos likewise has an easy solution. The iron bob swinging above the magnet is simply not a classical pendulum. So not all swinging weights that we would ordinarily think of as being pendulums are in fact classical pendulums. That the simple generalization fails to be true, however, causes no philosophical discomfort. Rather, the discovery of its falsity represents a scientific advance. We have discovered a new kind of pendulum, a magnetically augmented pendulum, in which the force of gravity is supplemented by a magnetic force directed toward a point directly below the point of rest. Constructing a theoretical model that does apply to such systems is a fairly easy problem in physics.

The philosophical price of this easy solution is that we give up the Aristotelian ideal that the empirical content of science is to be recorded in universal generalizations. Instead we have restricted generalizations whose scope may be well established. This seems far preferable to retaining generalizations that include indefinite provisos that may not even be formulatable in terms of current theory. Moreover, the suggested view, I would claim, provides a much more faithful account of relationships among laws, theories, and generalizations in science as it is practiced in the latter part of the twentieth century. But this is an empirical claim requiring evidence that I cannot now provide.

Conclusion

Our solution to the problem of provisos nicely illustrates the possibility of rejecting a major component of logical empiricism without thereby abandoning a "formal" or "analytical" approach to the philosophy of science. The problem is not solved by globalizing theories into paradigms or research traditions. Indeed, just the opposite. Here the basic form of a theoretical hypothesis is the attribution of a formally defined structure to an individual real system. Theories are then understood as being restricted sets of such basic hypotheses. There are, however, few restrictions on the internal complexity of a theoretical model. The suggested framework is therefore both more modest and more ambitious than logical empiricism. It abandons the Aristotelian ideal of looking for universal generalizations. But it gives full freedom to the construction of ever more complex theoretical models in the attempt to capture the structure of the various types of systems that make up our world. Such a framework should provide the basis for a more accurate and comprehensive theory of science than that suggested by either logical empiricism or its recent critics.

Notes

The author's research has been supported in part by a grant from the National Science Foundation.

1. The interpretation of recent philosophy of science expressed in this paragraph has already achieved textbook status, as, for example, in Harold Brown's *Perception, Theory and Commitment: The New Philosophy of Science* (Chicago: University of Chicago Press, 1977).

2. Support for this claim can be found in Peter D. Asquith and Henry E. Kuburg Jr., eds., *Current Research in the Philosophy of Science* (East Lansing, Mich.: Philosophy of Science Association, 1979). Several widely acclaimed recent books represent this continuing analytical tradition, including Bas C. van Fraassen's *The Scientific Image* (New York: Oxford University Press, 1980), and Clark Glymour's *Theory and Evidence* (Princeton, N.J.: Princeton University Press, 1980).

3. The recent literature on these problems begins with John Canfield and Keith Lehrer's paper "A Note on Prediction and Deduction," *Philosophy of Science* 28 (1961):204–208. Their position was generally supported by Wolfgang Stegmüller in "Explanation, Prediction, Scientific Systematization and Non-Explanatory Information," Ratio 8 (1966):1–24. The Canfield and Lehrer position was criticized by J. Alberto Coffa in "Discussion: Deductive Predictions," *Philosophy of Science* 35 (1968):279–283. Hempel took up the general issue in his unpublished 1970 Carus Lectures, "Problems and Changes in the Empiricist Conception of Theories," delivered at the Central Division meeting of the American Philosophical Association in St. Louis. His interest in the problem, as evidenced by recent lectures, continues undiminished.

4. Coffa argues that provisos, which he calls "extremal clauses," can be distinguished from general *ceteris paribus* conditions in a manner that prevents laws from becoming vacuous. These arguments are developed in his Ph.D. dissertation, "Foundations of Inductive Explanation" (University of Pittsburgh, 1973; Ann Arbor, Mich.: University Microfilms, 1973).

5. Patrick Suppes employed this point of view in his *Introduction to Logic* (New York: Van Nostrand, 1957). For a later, introductory exposition, see his "What Is a Scientific Theory?" in S. Morgenbesser, ed., *Philosophy of Science Today* (New York: Basic Books, 1967), pp. 55–67. J. D. Sneed's development of the approach appears in *The Logical Structure*

of *Mathematical Physics* (Dordrecht: Reidel, 1971). Wolfgang Stegmüller develops Sneed's position in *The Structure and Dynamics of Theories* (New York: Springer, 1976). Bas van Fraassen's version is presented in "On the Extension of Beth's Semantics of Physical Theories," *Philosophy of Science* 37 (1970):325–339, and in his *The Scientific Image.*

6. The problem of finding something to be the reference of the often honorific title "theory" is more one of terminology than substance. In several previous papers and in the first edition of my elementary textbook, *Understanding Scientific Reasoning* (New York: Holt, Rinehart and Winston, 1979), I identified theories with either sets of theoretical models of their linguistic formulations. In this paper, as well as in the second edition of the textbook (1984), I identify theories with sets of theoretical hypotheses. In a new book, *Explaining Science: A Cognitive Approach* (Chicago: University of Chicago Press, 1988), I suggest identifying theories as heterogeneous entities consisting of both sets of models and sets of theoretical hypotheses. Chap. 3 of this latter work expands on the approach of the present paper within a broader, "cognitive" framework for explaining the operations of modern science.

3. Rational Prediction

Wesley C. Salmon

A colleague, to whom I shall refer (quite accurately) as "the friendly physicist," recently recounted the following incident. While awaiting takeoff on an airplane, he noticed a young boy sitting across the aisle holding onto a string to which was attached a helium-filled balloon. He endeavored to pique the child's curiosity. "If you keep holding the string just as you are now," he asked, "what do you think the balloon will do when the airplane accelerates before takeoff?" The question obviously had not crossed the youngster's mind before that moment, but after giving it a little thought, he expressed the opinion that the balloon would move toward the back of the cabin. "I don't think so," said the friendly physicist; "I think it will move forward." The child was now eager to see what would happen when the plane began to move. Several adults in the vicinity were, however, skeptical about the physicist's prediction; in fact, a stewardess offered to wager a miniature bottle of Scotch that he was mistaken. The friendly physicist was not unwilling and the bet was made. In due course, the airplane began to accelerate, and the balloon moved toward the front of the cabin. The child's curiosity was satisfied;[1] the theory—that all objects that are free to move will move toward the back of the cabin when the plane accelerates—was falsified; and the friendly physicist enjoyed a free drink.

I have related this anecdote to point out that there are at least three—probably more—legitimate reasons for making predictions. First, we are sometimes curious about future happenings, and we want to satisfy that curiosity without waiting for the events in

Reprinted, with minor revisions, from *The British Journal for the Philosophy of Science* 32 (1981):115–125.

question to transpire. To do so, we may make wild guesses, we may employ superstitious methods of prediction, we may appeal to common sense, or we may use more sophisticated scientific theories. Second, we sometimes make predictions for the sake of testing a theory. In the example at hand, the prediction regarding the motion of the balloon was a rather good test of the hypothesis that all objects free to move in the cabin will tend to move toward the rear when the airplane accelerates. The fact that objects heavier than air tend to fall toward the earth when they are unsupported, while objects lighter than air (such as helium-filled balloons) tend to move in the opposite direction, suggests that the behavior of a helium-filled balloon has a reasonable chance of falsifying the hypothesis about the behavior of all material objects in the air-filled cabin of the accelerating airplane, if it is indeed false. Third, we sometimes find ourselves in situations in which some practical action is required, and the choice of an optimal decision depends upon predicting future occurrences. Although wagering is by no means the only such type of practical decision making, it is a clear and comprehensible example. We all agree, I take it, that scientific theories often provide sound bases for practical prediction.

A central feature of Sir Karl Popper's philosophy is his thesis concerning the status of induction. Indeed, he begins his book *Objective Knowledge* with the statement: "I think that I have solved a major philosophical problem: the problem of induction. . . . This solution has been extremely fruitful, and it has enabled me to solve a good number of other philosophical problems" (1972, p. 1). His solution, as is well known, involves a complete rejection of induction. This claim has been advanced in many of his writings spanning several decades, and it is reiterated in his autobiography (1974*a*) and in his "Replies to My Critics" (1974*b*).

For some time it has seemed to me that the crucial test of an anti-inductivist philosophy of science would be its capacity to deal with the predictive aspects of scientific knowledge. In a paper (Salmon 1968*a*) presented at the 1965 International Colloquium on Philosophy of Science at Bedford College, London, I attempted to offer a severe challenge to Popper's views concerning induction by posing what I took to be a serious dilemma: On Popper's account, either science embodies essential inductive aspects or else science is lacking in predictive content.[2] In the published proceedings of the Bedford College Colloquium (Lakatos 1968), J. W. N. Watkins contributed an answer to my critique. He denied that scientific reasoning is inductively infected, and he argued that it can, nevertheless, provide a basis for rational prediction. In Popper's replies to his critics (1974b, pp. 1028–1030), he acknowledges that I have understood his views "fairly well," and he endorses Watkins's response. I take this as evidence that we have located a genuine disagreement—one that is reasonably free from purely verbal disputes or out-and-out misrepresentations—regarding Popper's anti-inductivist stand. The question involves what Popper calls "the pragmatic problem of induction." It is this issue that I want to pursue in the present paper; it concerns the problem of rational prediction. Although the issue may appear to be rather narrow, it seems to me to have pivotal importance with regard to the assessment of Popper's deductivism.

Let me attempt to formulate the basic difficulty as I see it. In its very simplest terms, Popper's account of scientific knowledge involves generalizations and their observational tests. If we find a *bona fide* counterexample to a generalization, we can say that it has been deductively refuted. To be sure, as Popper explicitly acknowledges, there may be difficulties in some cases in determining whether certain observations constitute genuine counterexamples to a generalization, but that does not undermine the claim that a genuine counterexample yields a deductive refutation. According to Popper, negative instances provide rational grounds for rejecting generalizations. If, however, we make observations and perform tests, but no negative instance is found, all we can say deductively is that the generalization in question has not been refuted. In particular, positive instances do not provide confirmation or inductive support for any such unrefuted generalization. At this stage, I claim, we have no basis for rational prediction. Taken in themselves, our observation reports refer to past events, and consequently they have no predictive content. They say nothing about future events. If, however, we take a general statement as a premise, and conjoin to it some appropriate observation statements about past or present events, we may be able to deduce a conclusion that says something about future occurrences and that, thereby, has predictive content. Popper himself gives this account of *the logic of prediction* (1974b, p. 1030).

The problem of rational prediction concerns the status of the general premise in such an argument. One may claim, as Popper does, that we ought not to use a generalization that has actually been refuted as a premise in a predictive argument of this sort, for we are justified in regarding it as false. We ought not to employ premises that are known to be false if we hope to deduce true predictions. The exclusion of refuted generalizations does not, however, tell us what general premise should be employed. Typically there will be an infinite array of generalizations that are compatible with the available observational evidence, and that are therefore, as yet, unrefuted. If we were free to choose arbitrarily from among all the unrefuted alternatives, we could predict anything whatever. If there were no rational basis for choosing from among all of the unrefuted alternatives, then, as I think Popper would agree, there would be no such thing as rational prediction. We are not in this unfortunate situation, Popper contends, for we do have grounds for preferring one unrefuted generalization to another: "My *solution* of the logical problem of induction was that we may have *preferences* for certain of the competing conjectures; that is, for those which are highly informative and which so far have stood up to eliminative criticism" (1974*b*, p. 1024). Popper's concept of corroboration is designed to measure the manner in which conjectures have stood up to severe criticism, including severe testing. This, I take it, is the crucial thesis—that *there is a rational basis for preferring one unrefuted generalization to another for use in a predictive argument*. If that is correct, then Popper can legitimately claim to have solved the problem of rational prediction.

If we are going to talk about preference among generalizations, then we have to be quite explicit about the purpose for which the generalization is to be used. In this context, we are discussing prediction, so the preference must be in relation to predictive capability. As Popper rightly insists, any generalization we choose will have predictive import in the sense that it will make statements about future events—more precisely, in a predictive argument as characterized above, it yields conclusions about future occurrences. But since all of the various unrefuted generalizations have predictive content in that sense, we must still ask on what basis the predictive content of one conjecture is rationally preferable to that of another conjecture.

At this stage of the discussion, it is important to recall the point of the opening story, namely, that predictions are made for various purposes. Thus, even if we agree that we want to select a generalization for predictive purposes, we must still specify what type of prediction is involved. Popper explicitly acknowledges (1974*b*, pp. 1024–1025) that there are two types of preference, "the theoretician's preference" and that of "the man of practical action." As I understand Popper's view, the theoretician is interested in formulating bold conjectures that have high content and in subjecting them to severe tests. Insofar as the theoretician is mainly interested in explanations of known phenomena, he may not be much involved in making any sorts of predictions. I suppose we might distinguish the theoretician's explanatory preference from the theoretician's predictive preference, recognizing that there is bound to be a close connection between preferences of these two kinds. When the theoretician is actually involved *qua theoretician* in making predictions, the purpose is to devise (and, perhaps, to instruct the experimentalist on how to conduct) a severe test. The purpose of predictions made in this theoretical context is to gain information that is useful in the evaluation of scientific theories. If the chief value of the scientific theories is explanatory, then it is not at all clear that a primary desideratum of the predictive argument is to arrive at a true prediction. As Popper has emphasized, and as all of us know, a false prediction can be valuable, since the realization (on the basis of observation) that it is false can be highly informative.

Having briefly characterized theoretical preference, let us now focus attention upon the kind of preference that is pertinent to the practical context, with special attention to the kinds of predictions that play a role in practical decision making. As I have remarked above, Popper claims that for theoretical purposes we prefer theories that are highly corroborated to those that are less well corroborated. I do not think this claim is unproblematic, but I do not propose arguing the matter here. My aim is to emphasize that, even if we are entirely justified in letting such considerations determine our theoretical preferences, it is by no means obvious that we are justified in using them as the basis for our preferences among generalizations that are to be used for prediction in the practical decision-making context. Popper and Watkins have maintained, however, that corroboration should play a crucial role in determining both theoretical preference and practical preference.

Since scientific theories are used for both theoretical and practical purposes—including prediction—and since, according to Popper, theory preference is based upon corroboration, I had mistakenly inferred (prior to 1968) that the appraisal of a theory in terms of corroboration must imply some attempt at an appraisal of the theory with respect to its future performance. If that were Popper's thesis, I had argued, then corroboration must involve some element of induction (or nondemonstrative inference of some sort), for past performance of the theory is taken to constitute a basis for some sort of claim about future performance. However, I have since been informed by Watkins (1968) and Popper (1974a) that I had misconstrued Popper's view. Statements about the corroboration of theories are no more than appraisals of their past performances; corroboration statements hold no predictions with respect to future performance. If they did, they would be inductive (as I had claimed); but they are not inductive, so they cannot be predictive.

This view of corroboration holds serious difficulties. Watkins and Popper agree, I take it, that statements that report observations of past and present events do not, in and of themselves, have any predictive content. Moreover, they maintain, statements about the corroboration of conjectures do not, in and of themselves, have any predictive content. Conjectures, hypotheses, theories, generalizations—call them what you will—do have predictive content. The problem is that there are many such general statements, rich in predictive content, that make incompatible predictive claims when conjoined with true statements about past and present occurrences. The fact that a general statement has predictive content does not mean that what it says is true. In order to make a prediction, one must choose a conjecture that has predictive content to serve as a premise in a predictive argument. In order to make a *rational* prediction, it seems to me, one must make a *rational* choice of a premise for such an argument. But from our observational evidence and from the statements about the corroboration of a given conjecture, no predictive appraisal follows. Given two conjectures that, in a particular situation, will lead to incompatible predictions, and given the corroboration ratings of these two hypotheses, *nothing follows* about their comparative predictive capacities. Thus, it seems to me, corroboration—the ground for theoretical preference—furnishes no rational basis for the preference of one conjecture to another *for purposes of practical prediction.* I am not complaining that we are not told *for sure* that one will make a correct prediction and that the other will not. I am complaining that no rational basis whatever has been furnished for a preference of *this* type.

In his reply to my Bedford College paper, Watkins acknowledges that there is an important distinction between theoretical and practical preferences, and he further acknowledges that the two kinds of appraisal may have quite different bases:

> Now our methods of hypothesis-selection in practical life should be well suited to our practical aims, just as our methods of hypothesis-selection in theoretical science should be well suited to our theoretical aims; and the two kinds of method may very well yield different answers in a particular case. (1968, p. 65)

He goes on to explain quite correctly how utility considerations may bear upon the practical situation. Then he considers the case in which utility does not play a decisive role:

> Now suppose that, for a particular agent, the mutually incompatible hypotheses h_1 and h_2 are on a par utility-wise, and that in the situation in which he finds himself, he has *got* to act since 'inaction' would itself be one mode of action. Then if h_1 is the only alternative to h_2 before him, he *has* to choose one of them. Then it would be rational for him to choose the better corroborated one, the one which has withstood the more severe criticism, since he has nothing else to go on. (Pp. 65–66)

Watkins offers no further argument for supposing that corroboration provides a rational basis for *practical* preference. Moreover, the hint of an argument which he does supply appeals to a false premise. The agent does have other things "to go on." He could decide between the two hypotheses by the flip of a coin. He could count the numbers of characters in each of the two hypotheses in the particular formulation given, and choose the one that has fewer. He could choose the hypothesis that comes first lexicographically in the given formulation. What Watkins is suggesting, it seems to me, is not that the agent has "nothing else to go on" but rather that he has no other *rational* basis for preference. But such an argument would be patently question begging. Even if all other bases for choice were irrational, it would not follow that the one cited by Watkins is *ipso facto* rational. Indeed, if we take seriously Popper's statement, "I regarded (and I still regard) the degree of corroboration of a theory merely as a critical report on the quality of past performance: *it could not be used to predict future performance*" (1974a, p. 82), it is hard to see how corroboration can supply a rational basis for preference of a theory *for purposes of practical prediction*.

Whether my criticism of Popper's position is correct or incorrect, the issue I am raising has fundamental importance. For if it should turn out that Popper could not provide a tenable account of rational prediction, then—given his persistent emphasis upon objectivity and rationality—we could hardly credit his claim to have solved the problem of induction. Moreover, in his replies to his critics, Popper acknowledges the issue. With the comment, "Our corroboration statements have no predictive import, although they motivate and justify our *preference* for some theory over another" (1974*b*, pp. 1029–1030), he endorses the answer Watkins had furnished. Since I am not attempting to deal with the psychological problem of induction, I shall not dispute the claim that corroboration may *motivate* the preference of one theory to another. What I want to see is how corroboration could *justify* such a preference. Unless we can find a satisfactory answer to that question, it appears to me that we have no viable theory of *rational* prediction, and no adequate solution to the problem of induction.

In *Objective Knowledge*, Popper offers an answer to the basic question which seems closely related to that of Watkins:

> [A] *pragmatic belief in the results of science* is not irrational, because there is nothing more 'rational' than the method of critical discussion, which is the method of science. And although it would be irrational to accept any of its results as certain, there is nothing 'better' when it comes to practical action: there is no alternative method which might be said to be more rational. (1972, p. 27)

This response appears to miss the point. The question is not whether other methods—for example, astrology or numerology—provide more rational approaches to prediction than does the scientific method. The question is whether the scientific approach provides a more rational basis for prediction for purposes of practical action than do these other methods. The position of the Humean skeptic would be, I should think, that none of these methods can be shown either more or less rational than any of the others. But if every method is equally lacking in rational justification, then there is no method that can be said to furnish a rational basis for prediction, for any prediction will be just as unfounded rationally as any other. If the Humean skeptic were right, we could offer the following parallel claim. A pragmatic belief in the predictions found in Chinese fortune cookies is not irrational, for there is nothing more rational . . .

In his replies to his critics, Popper again addressed the problem, and he came more firmly to grips with it:

> But every action presupposes a set of expectations, that is, of theories about the world. Which theory shall the man of action choose? Is there such a thing as a *rational choice*?
>
> This leads us to the *pragmatic problems of induction*, which to start with, we might formulate thus:
>
> (*a*) Upon which theory should we rely for practical action, from a rational point of view?
>
> (*b*) Which theory should we prefer for practical action, from a rational point of view?
>
> My answer to (*a*) is: from a rational point of view, we should not 'rely' on any theory, for no theory has been shown to be true, or can be shown to be true (or 'reliable').
>
> My answer to (*b*) is: we should *prefer* the best tested theory as a basis for action.
>
> In other words, there is no 'absolute reliance'; but since we *have* to choose, it will be 'rational' to choose the best tested theory. This will be 'rational' in the most obvious sense of the word known to me: the best tested theory is the one which, in the light of our *critical discussion*, appears to be the best so far; and I do not know of anything more 'rational' than a well-conducted critical discussion. (1974*b*, p. 1025)

Let us not be seduced by honeyed words. If we wish to claim that a theory "appears to be the best so far," we must ask, "Best for what purpose—theoretical explanation or practical prediction?" Since it is "the best tested theory" and it has been subjected to "critical discussion," then, in the light of the many statements by Popper and others about the lack of predictive import of corroboration, we must conclude, I believe, that the answer is, "Best for theoretical explanation." Perhaps I am being unduly obtuse, but I cannot see that any reason has been provided for supposing that such a theory is best *for practical prediction*.

I must confess to the feeling that we have been given the runaround. We begin by asking how science can possibly do without induction. We are told that the aim of science is to arrive at the best explanatory theories we can find. When we ask how to tell whether one theory is better than another, we are told that it depends upon their comparative ability to stand up to severe testing and critical discussion. When we ask whether this mode of evaluation does not contain some inductive aspect, we are assured that the evaluation is made wholly in terms of their comparative success up to now; but since this evaluation is made entirely in terms of past performance, it escapes inductive contamination because it lacks predictive import. When we then ask how to select theories for purposes of rational prediction, we are told that we should prefer the theory that is "best tested" and that "in the light of our *critical discussion*, appears to be the best so far," even though we have been explicitly assured that testing and critical discussion have no predictive import. Popper tells us, "I do not know of anything more 'rational' than a well-conducted critical discussion." I fail to see how it could be rational to judge theories *for purposes of prediction* in terms of a criterion that is emphatically claimed to be lacking in predictive import.[3]

Fearing that the point of his initial argument may have been missed, Popper attempts another formulation:

> Let us forget momentarily about what theories we 'use' or 'choose' or 'base our practical actions on', and consider only the resulting *proposal* or *decision* (to do X; not to do X; to do nothing; or so on). Such a proposal can, we hope, be rationally criticized; and if we are rational agents we will want it to survive, if possible, the most testing criticism we can muster. *But such criticism will freely make use of the best tested scientific theories in our possession.* Consequently any proposal that ignores these theories (where they are relevant, I need hardly add) will collapse under criticism. Should any proposal remain, it will be rational to adopt it.
>
> This seems to me all far from tautological. Indeed, it might well be challenged by challenging the italicized sentence in the last paragraph. Why, it might be asked, does rational criticism make use of the best tested although highly unreliable theories? The answer, however, is exactly the same as before. Deciding to criticize a practical proposal from the standpoint of modern medicine (rather than, say, in phrenological terms) is itself a kind of 'practical' decision (anyway it may have practical consequences). Thus the rational decision is always: adopt critical methods which have themselves withstood severe criticism. (1974b, pp. 1025—1026)

I have quoted Popper *in extenso* to try to be quite sure not to misunderstand his answer. The italicized sentence in the first paragraph raises precisely the question that seems to me crucial. In the second paragraph, Popper admits the legitimacy of the question, and he offers an answer. When he says, "The answer . . . is exactly the same as before. . . . [T]he rational decision is always: adopt critical methods which have themselves withstood severe criticism," he seems to be saying that we should adopt his methodological recommendations, because they have "withstood severe criticism." But his answer is inappropriate in this context because our aim is precisely to subject his philosophical views, in the best Popperian spirit, to severe criticism.

In my reply to Watkins, I said, "Watkins acknowledges . . . that corroboration does have predictive import in practical decision making" (1968*b*, p. 97). Popper has objected to this way of putting the matter: "[O]*ur theories do have predictive import*. Our corroboration statements have no predictive import, although they motivate and justify our *preference* for some theory or other" (1974*b*, pp. 1029–1030). Let us grant that corroboration statements have no predictive *content*—indeed, that they are analytic, as Watkins remarks (1968, p. 63)—and that theories are the kinds of statements that do have predictive *content*. It does not follow, as Popper has claimed, that corroboration has no predictive *import*. The distinction between predictive content and predictive import is no mere verbal quibble; a fundamental substantive point is at issue. Statements whose consequences refer to future occurrences may be said to have predictive content; rules, imperatives, and directives are totally lacking in predictive content because they do not entail any statements at all. Nevertheless, an imperative—such as "No smoking, please"—may have considerable predictive import, for it may effectively achieve the goal of preventing the occurrence of smoking in a particular room in the immediate future.

Since corroboration, in some cases at least, provides the basis for deciding which theory (with its predictive content) is to be used for the purpose of making practical predictions, it seems to me that corroboration, even if it is lacking in predictive content, does have enormous predictive import. Perhaps this point can be put more clearly in the following way. *Statements* assessing the corroboration of theories have no predictive *content*, as Popper, Watkins, and others maintain. The *directive*—to choose more highly corroborated theories in preference to theories that are less well corroborated for purposes of practical prediction—has considerable predictive *import*. The problem, which it seems to me the anti-inductivists have failed to solve, is how to vindicate this directive for making predictions.[4] Without some sort of vindication for this directive, the problem of rational prediction remains unresolved.

I have wondered why it would seem evident to Popper that corroboration, as he construes it, should provide a guide to rational prediction. In his autobiography, he gives what appear to be indications of an answer.

> I regarded (and I still regard) the degree of corroboration of a theory merely as a critical report on the quality of past performance: *it could not be used to predict future performance.* . . . When faced with the *need to act*, on one theory or another, the rational choice was to act on that theory—if there was one—which so far had stood up to criticism better than its competitors had: there is no better idea of rationality than that of a readiness to accept criticism. Accordingly, the degree of corroboration of a theory was a rational guide to practice. (1974*a*, p. 82)

A further elaboration of the theme informs us that

> when we think we have found an approximation to the truth in the form of a scientific theory which has stood up to criticism and to tests better than its competitors, we shall, as realists, accept it as a basis for practical action, simply because we have nothing better (or nearer to the truth). (ibid., pp. 120–121)

Realism is a position to which Popper has adhered since the time of his earliest philosophical activity; near the beginning of his autobiography he tells us that "a realist who believes in an 'external world' necessarily believes in the existence of a cosmos rather than a chaos; that is, in regularities" (ibid., p. 14). Thus, I am led to conjecture, it may be that Popper's adherence to the thesis that corroboration can provide a basis for rational prediction rests ultimately upon his realism, which embodies a version of a principle of uniformity of nature. If this suggestion is correct, we can still legitimately wonder whether Popper's epistemology is as far from traditional inductivism as he would have us believe.

To conclude this discussion, I should like to recall the point of my opening anecdote. It seems to me incorrect to suppose that the only concern of *theoretical* science is to make bold explanatory conjectures that can be tested and criticized. It is a mistake, I believe, to suppose that all prediction, aside from that involved in the testing of theories, is confined to contexts in which practical action is at stake. Theoretical science furnishes both explanations and predictions. Some of these predictions have practical consequences and others do not. When, for example, scientists assembled that there was a grain of truth in these views, but I am here more concerned with the deluge of dross than the grains of truth. Put in a more kindly way, I would suggest that the deductivist view presents an idealization that is fundamentally misleading in a number of important respects.

Defects of Deductivism

None of the shortcomings of the deductive model that I shall mention here are particularly novel or surprising. Most have been considered by deductivists, and dismissed by a handwaving reference to idealization. Nevertheless it is worth reviewing them briefly, since their collective import bodes ill for the deductivist core characterized above.

Let us try to be a bit more precise about what we are trying to do. Consider some body of scientific knowledge. Let us leave open the question what this body of knowledge contains. But let us suppose that there is a hypothesis that is said to be acceptable or confirmed by holders of this body of scientific knowledge. As philosophers, we take the holders of this body of knowledge at their word—or at any rate we are careful not to choose controversial examples. What we want to do is to provide a rational reconstruction of the relation between the body of knowledge and the hypothesis H in virtue of which we can show not only that H is confirmed or acceptable but also *why* it is confirmed or acceptable. We wish to understand what is going on. Now of course it *may* be that if we *really* understood what was going on, we would understand that no rational reconstruction is possible—that scientific inference is essentially irrational and idiosyncratic. Such a negative existential claim (in a couple of senses) would be very hard to support, and most philosophers, I think, would not find it very appealing. We therefore suppose that some sort of rational reconstruction is possible, and we set about trying to discover it or invent it.

A natural first step is to reconstrue the vaguely understood "body of knowledge" as something about which it is possible to be more explicit. We may consider it to be a set of sentences in a first-order language; a set of sentences in a first-order language to which have been added axioms of set theory; a set of sentences in a higher-order language; a set of sentences in a modal language; or a set of models of one of these sets of sentences.

Whatever the ingredients of this set of statements, it is assumed to be *deductively closed*; that is, it is assumed to contain the consequences of any finite subset of its members. If it is not to be totally useless, therefore, it must be assumed to be consistent, else it would contain all statements of our language whatever. Thus the deductivist assumes a deductively closed and consistent set of statements that can serve as premises for deductions; generally this set will include theories and generalizations (of whatever provenance), as well as boundary conditions and other observationally warranted statements. It may also include, of course, analytic statements, bridge principles, and the like that may or may not be construed as having observational content.

Fourth, against this background, scientific inquiry is supposed to proceed in roughly this way: A hypothesis H is up for test or examination. H, together with a set of theoretical background statements, mixed statements, boundary conditions in the observational vocabulary, etc., entails (logically implies) some observation sentence O. To put the hypothesis to the test, we set up the experimental conditions (thereby putting ourselves in a position to accept all the premises with the possible exception of H itself), examine the world, and, as the case may be, observe O or observe $\sim O$. In the latter case, we have, *ceteris paribus, refuted H*; in the former we may have *confirmed H*—at least if O is an instance of it—or anyway we will have "failed to refute" H.

Note that this very general description includes classical statistical testing: The background statements contain information about known causal connections, our sampling procedures, and the like; H is taken as a null hypothesis; our boundary conditions include a description of our sampling procedures as actually applied; and we *deduce* from H, our background information, and our proposed sampling procedures that the probability that the sample will have such and such a characteristic is (say) .05. When we have taken our sample or performed our experiment, and have discovered that the sample has the characteristic in question, we reject the null hypothesis H at the .05 level; or if the sample lacks that characteristic, rather than saying that the null hypothesis is confirmed, we just keep our mouths shut.

What could possibly be wrong with such an innocuous description of scientific method as this, and how could it possibly be in need of justification? Of course, everybody admits that it is an idealization in a number of respects. It would be irrational to deny

Given such a wide variety of deductivist approaches to scientific inference, it is hard to characterize deductivism in much detail. But there is a core to the deductivist model that represents, to a first approximation, what most of those philosophers I have mentioned seem to be committed to. First of all, the deductivist is committed to the existence of an *observational vocabulary*, in which the observational or testable deductive consequences of theories, laws, and hypotheses can be expressed. Correlatively, since it is obvious that not all of the statements and predicates of a body of scientific knowledge can be expressed in an observational vocabulary, the deductivist is committed to a distinction in kind between an observational and a *nonobservational vocabulary*. This is not to say that the distinction must be regarded as ironclad or as a priori; it may be context dependent (in some sense), historically relative, or what have you. The point is rather that in a certain context, or at a certain epoch, there is a set of statements that function like the *protocol sätze* of positivism: they are the sorts of statements that are tested directly by observation. While we may argue about when Herschel observed Uranus, we do not argue about the fact that on March 13, 1781, he observed "a curious . . . nebulous star or perhaps a comet"—that is, a luminosity of a certain character at a certain position (Kuhn 1977, p. 171).

Second, it is universally agreed among deductivists that scientific theories, laws, and hypotheses have *empirical* content. Some writers take the empirical content of these scientific generalizations to be constituted by the set of observation statments they imply; others construe "empirical content" more broadly. For my purposes, the distinction between these views is irrelevant. What I wish to call attention to is that, on the deductivist view, the generalizations of science have as deductive consequences certain statements expressible in whatever is taken to be, in a given context or at a given moment, the observational vocabulary.

Third, deductivists agree that the scientist accepts a set of statements that are allowed to function as premises in deductive arguments. This set of statements certainly includes the observation statements that are warranted by experience. In all but a very few cases, it also includes some statements of a more general or theoretical nature. Among these may be previously accepted theories and generalizations that count as background knowledge; very general and abstract statements characterizing a research program; and theories relative to which other laws and theories may be confirmed.

often said, you have to begin with a simplification. Yet it seems clear that the motivation for his work stemmed in part from the conviction that scientific hypotheses could be confirmed—made more credible—by the successful testing of their deductive observational consequences.

It is more curious yet that many philosophers and statisticians who have been concerned with statistical inference, and indeed the vast majority of those who employ statistical inference as an everyday scientific tool, accept the deductive model. Significance testing falls right into the classical pattern: The scientist forms a clear null hypothesis; from that hypothesis, together with auxiliary assumptions regarding sampling procedures, a statistical consequence is deduced. This consequence has the general form that a certain sort of observation is very rare, or very improbable. The scientist performs the experiments or takes the samples, and makes precisely the sort of observation whose rarity follows from the null hypothesis. He therefore rejects the null hypothesis. The analogy to *modus tollens* is clear.

Even such careful and nondogmatic empiricists as Ernest Nagel (1961) and Carl G. Hempel (1965) accept the general deductivist picture. They reject any simpleminded formula for determining the empirical content of a scientific law or hypothesis; they credit background knowledge and auxiliary hypotheses with playing an essential role in scientific inquiry. Nevertheless, they still suppose that deduction from scientific hypotheses (together with background knowledge, auxiliary hypotheses, and boundary conditions) can yield empirically testable results; that under some circumstances, which may be difficult to specify formally, the refutation of such results by observational test can cast doubt on these hypotheses; and that under some circumstances, which again may be difficult to specify formally, the verification of such results by observational test can lend support to these hypotheses.

The most recent version of deductivism is that of Clark Glymour (1980). He argues that scientific hypotheses can (as we thought all along?) be confirmed by their instances, but only relative to a general theory. The logic involved in testing a law or hypothesis is, as usual, straightforwardly deductive. the novelties of his approach have to do with the way in which confirmation is relativized to a theory, and with the fact that (by the bootstrap technique) a theory can be confirmed relative to itself.

daring hypotheses. The measure of the power and daring of a hypothesis lies in its testable consequences. The essence of science is the endeavor to refute the daring hypothesis by seeking a deductive consequence that can be falsified by experience. Most writers do not generally regard things as being quite this simple, however. Boundary conditions and auxiliary hypotheses are often required for the deduction of testable consequences from hypotheses. While a simple universal generalization can be refuted by a counterinstance (by *modus tollens*), a hypothesis with mixed universal and existential quantifiers cannot be so refuted.

Imre Lakatos (1970), recognizing the importance of auxiliary hypotheses and boundary conditions, supposes that the falsification criterion can only be applied within what he calls a *research program*. A research program provides a bundle of assumptions and commitments within which certain kinds of limited hypotheses can serve to generate testable consequences that can be refuted (or confirmed) by observation. The logic, however, is still deductive; the rule of inference applied within the program is still essentially *modus tollens*.

Thomas Kuhn (1962), Paul Feyerabend (1970), and others who are more interested in history (of a certain sort) than formal reconstruction recognize that scientific hypotheses are rarely *refuted* until long after they are dead. While the hypothesis still has its teeth, a "refuted" consequence is a reflection on the experimenter and not on the hypothesis. But the internal structure is still deductive. The consequences we draw from our theories, combined with observational data, are still deductive consequences. However, they are no longer, as they were for Popper and his followers, experimental consequences to be put to observational test with a view to assessing the worth of the theories. Rather, they are consequences that in turn, combined with more observational data, yield the conclusions that constitute the body of scientific doctrine accepted at a given time.

What is more curious is that the great modern upholder of inductive logic, Rudolf Carnap (1963), as well as others who have followed him in focusing on confirmation, is also a deductivist. He never questioned the claim that scientific hypotheses could be refuted by contrary instances, and rather focused on the question of the support given them by positive instances. His own systems did not get very far in elucidating the scientific process of confirmation, since they were designed for very elementary languages. As Carnap

4. The Justification of Deduction in Science

Henry E. Kyburg, Jr.

What I am concerned about here is the deductive model of scientific inquiry, construed very loosely and generally. I shall argue that there is a core of agreement among the proponents of the deductive model; and that this core of agreement is perniciously misleading, even when one takes account of the fact that it is intended as an idealization. I shall briefly characterize an alternative model, which for want of a better contrasting term I shall call the *inductive* model. Then I shall argue that this inductive model captures the useful and realistic features of the deductive one, and further that it illuminates certain aspects of scientific inquiry that are obscured by deductivism.

The Deductive Model

The hypothetico-deductive model of scientific method has been around for a long time. It exists in various forms, and in various degrees of sophistication within a given form. One is still exhorted—particularly in courses in methodology associated with the social and behavioral sciences—to formulate one's hypotheses clearly, to draw testable implications from them by deductively valid reasoning, and to put these implications to experimental test without fear or favor. This is all very sound advice, no doubt. But in this very general form, it doesn't provide us with much insight into the workings of scientific inference. The moment we begin to put flesh on these bones, we encounter controversies and problems.

Karl Popper's approach to scientific inference falls into this pattern. According to Popper (1959), we are to invent powerful and the first man-made atomic pile under the West Stands at the University of Chicago, they had to make a prediction whether the nuclear chain reaction they initiated could be controlled, or whether it would spread to surrounding materials and engulf the entire city—and perhaps the whole earth—in a nuclear holocaust. Their predictions had both theoretical and practical interest. Contemporary cosmologists, for another example, would like to *explain* certain features of our universe in terms of its origin in a "big bang"; many of them are trying to *predict* whether it will end in a "big crunch." In this case, the predictive question seems motivated by pure intellectual curiosity, quite unattached to concerns regarding practical decision making. Whether a helium-filled balloon will move forward in the cabin of an airplane when the airplane accelerates, whether a nuclear chain reaction—once initiated—will run out of control, and whether the universe will eventually return to a state of high density are all matters of legitimate scientific concern.

In this paper, I have attempted to argue that pure deductivism could not do justice to the problem of rational prediction in contexts of practical decision making. If we ask whether Popperian deductivism can adequately account for scientific predictions of the more theoretical varieties, then I suspect that we would have to go through all of the preceding arguments once more. The net result would be, I think, that science is inevitably inductive in matters of intellectual curiosity as well as practical prediction. It *may* be possible to excise all inductive ingredients from science, but if the operation were successful, the patient (science), deprived of all predictive import, would die.

Notes

A version of this paper was presented orally at the Symposium on the Philosophy of Sir Karl Popper, London School of Economics, July 14–16, 1980. This material is based upon work supported by the National Science Foundation (U.S.A.) under Grant No. SES-7809146.

1. His curiosity regarding *what* would happen was satisfied, though not his curiosity as to *why*.

2. Similar themes were developed in Salmon 1967, chap. 2, sec. 3.

3. The argument advanced in this paragraph bears a strong resemblance, I think, to one developed in Grünbaum 1976; see esp. p. 246.

If we consider a relatively simple first-order language, we can define a measure function on it, and then a confirmation function, which will allow us to say that H has a high degree of confirmation relative to a suitable body of data, even when H is a universal generalization or a sentence with mixed quantifiers (Hintikka 1966). But when the language is rich enough to contain mathematics and scientific theories and hypotheses couched in mathematical terms (by which I mean to include any theories or hypotheses involving *quantities*), then it is difficult to see how measures can be assigned to all the statements of the language in such a way that hypotheses can come to be confirmed. A quantitative notion of confirmation does not seem to be globally applicable to the kinds of language in which ordinary science can be represented.

One is troubled, even in the case of relatively weak languages, by the arbitrary element in the specification of a logical measure—those Greek letter parameters (alpha, lambda, eta, etc.) that multiply as the language in question becomes more powerful. One might regard them as representing a part of the *characterization* of a language: specifying a language to represent a fragment of scientific discourse might be taken to require not only the specification of a set of predicates and operations but also the specification of the parameters determining the logical measure to be imposed on the sentences of the language.

But this introduces another formidable problem. The formulation of a scientific hypothesis is often accompanied by the introduction of new terminology or by the reinterpretation of old terminology. If this is construed as a change of language, then we must suppose that the logical measure function relevant to the evaluation of this new hypothesis will be that appropriate to the new language. But then how can we compare the degrees of evidential support of two hypotheses formulated in different languages? Profound incommensurability looms.

The question of acceptability is even more difficult. If H is accepted into our body of knowledge, it has a probability of one relative to that body of knowledge. But it cannot come to have a probability of one relative to a body of knowledge via conditionalization, unless it has a prior (a priori) probability of one. Isaac Levi (1981) has pointed out, however, that just as most Bayesians may allow the assignment of a probability of one to an observation statement under certain circumstances, so they are free to specify circumstances under which other statements may be assigned a probability of one. So far as I know only Levi has approached this problem of the "revision of knowledge" directly, however. We could eschew acceptance of scientific laws and hypotheses altogether, but this seems a heroic measure, and makes it difficult to understand the force of predictive and explanatory deductive arguments that employ such laws and hypotheses as premises.

Of course a deductivist need not be wedded to some form of probabilistic confirmation theory. He may settle for refutation or for some qualitative form of instance confirmation as a way of sorting goat hypotheses from sheep hypotheses. To discuss these alternatives, we must look more closely at the observation sentences.

What about that set of sentences we accept on the basis of experience, and relative to which we judge hypotheses? The traditional view (like most views cited as "traditional," it may never have been held) has it that this set of sentences contains protocol sentences, or observation sentences, which are to be regarded as unproblematic. It is well known that it is difficult to find a realistic way of specifying what they are: Are they sense-data statements? Or ordinary statements about ordinary-sized physical objects? Measurement statements? It has been pointed out (Kuhn 1977) that what counts as an "observation" at one point in historical time may not count as such at another. Even the "observation vocabulary" of a scientific discipline changes over time.

The very distinction between the "observation vocabulary" and the rest of the scientific vocabulary becomes more obscure the more we try to clarify it. It has been suggested (Hempel 1970) that the only distinction we can draw is between "antecedently understood" vocabulary and newly introduced vocabulary. But even this distinction is a slippery one, and it leaves untouched the problem that parts of the "antecedently understood" vocabulary may, through the development of new hypotheses, come to be regarded as antecedently *mis*understood. What was regarded as a perfectly sound observation sentence at one time may be regarded as theoretical or as meaningless at another. We must invoke context or history.

The idea behind having "observation" sentences as a distinguished part of our scientific vocabulary is that we want in our body of statements some (at least) of which we can be sure, and relative to which we can judge the credibility or acceptability of others. The most common view is that these are statements about middle-sized physical objects that are, under the appropriate circumstances, simply accepted and incorporated into our stock of scientific data. It is no doubt admitted that this is an idealization. Such statements can be erroneously accepted: a man may suffer from hallucination. But this posibility is not something, according to most deductivist writers, that we can or should take seriously. A little idealization never hurt anybody.

On the contrary, I think this little bit of idealization has had a pernicious effect on the philosophy of science. By focusing on strange and unusual sources of error, we have been led to ignore more usual and more typical kinds of error. And this has lent the deductive model, even in its probabilistic forms, more plausibility than it has deserved. In the conduct of science, error cannot be ignored. Many scientific laws and hypotheses are quantitative in form: they stipulate that certain mathematical relations hold among certain sets of quantities. Such laws, and the theories that imply them, are evaluated against measurements: the data consist of a series of measurements. But this renders the deductive model particularly puzzling, since nobody ever maintained that quantitative statements can be regarded as certain or incorrigible—they themselves are just the sort of statements we are alleged to accept or to confirm on the basis of more fundamental statements, for example, statments about the results of measurement.

Measurement

A somewhat closer look at measurement may prove enlightening. Traditional treatments of measurement suppose that we begin with an observable relation holding between middle-sized objects or processes: for example, "longer than." This is just the sort of predicate that is considered "observational" in deductivist reconstructions of science. It is supposed that we confirm inductively (or "fail to falsify") that this relation is transitive, and even that its relational complement is transitive; that is, that if a fails to bear R to b, and b fails to bear R to c, then a fails to bear R to c. It is this latter that is required for the characterization of the equivalence classes that will serve as the basis of a mathematical structure. But already we are in trouble, for not only are there difficulties with the confirmation—to the point of *acceptance*, be it noted—of general statements such as this but, in the case of the transitivity of the relational complement, we know that the generalization is frequently refuted. People not infrequently judge *a* to be the same length as *b*, and *b* to be the same length as *c*, but *c* not to be the same length as *a*.

The solution is obvious: distinguish between veridical judgments and erroneous ones. Of the three judgments mentioned above, at least one must be in error. If *a* is the same length as *b*, and *b* the same length as *c*, *c* *must* be the same length as *a*. But now "*a* is (is not) the same length as *c*" can no longer be regarded as a simple observation statement; the same is true of sentences of the form "*a* is longer than *b*."

If these statements are no longer to be regarded as observational, then (according to the deductive model) there must be some other statements relative to which they can become acceptable or confirmed. The natural candidates are observation *reports*: the reports of observations or the results of actual measurements. We recognize that these reports of the results of measurement may be in error—in fact, according to the usual treatment of errors of observation, they are almost certainly in error—but we have a quantitative theory of error that yields a probabilistic relation between the observations and the statements about true length relationships.

But how do we obtain this quantitative theory of error? We cannot obtain it from a statistical analysis of the relation between the reports and the realities, since the only access we have to the realities is by means of the reports. Furthermore, it is now clear that even the simplest example of the hypothetico-deductive method—the refutation of a hypothesis by an observed counterinstance—is now embedded in a morass of probabilistic reasoning. The theory of measurement itself is not only a proper theory; it is a theory that cannot be refuted in any simple way by a contrary observation, and this deductive awkwardness immediately spreads to the testing of any quantitative law.

Consider the law that under certain conditions, the pressure and volume of a sample of gas are inversely proportional. Leave to one side the widely acknowledged difficulty of "specifying" the "conditions." Given that we have the right conditions, all we need is a sample of gas x, such that $P(x)V(x)$ is not equal to $P'(x)V'(x)$. But what we have to go on are not the pressures and volumes but merely the results of measurements of those quantities. According to the usual theories of error, errors of arbitrarily large magnitude have non-zero probabilities. Thus a measurement of pressure cannot *rule out* any actual pressure, though (given the theory of error) it makes some values highly improbable. Skipping lightly over problems of statistical inference, direct inference, and inverse inference, we may suppose that, given our actual measurements, it is very *unlikely* that the product $P(x)V(x)$ is the same as the product $P'(x)V'(x)$. Is this a refutation of the proposed law? Not unless we construe refutation probabilistically. But the probabilistic analogue of *modus tollens*—from "Almost all A are B" and "a is not B" to infer "a is not A"—does not hold in general.

It is worth thinking generally about what happens when we learn to measure something. At a nonscientific and hypothetical Stage I, we have judgments of relative magnitude, which we might construe as protocol sentences. Obviously, these sentences are collectively inconsistent with the general axioms of structure we require to establish a quantitative theory of measurement. But we do not (as good deductivists should) abandon the possibility of measurement. Rather, we go on to Stage II: we assume that some of our judgments of relative magnitude are in error, and that conflict with the required axioms must be construed as *evidence of error* rather than as evidence that the necessary axioms are false.

The deductivist already has serious problems, for when we assume this we do not assume—and it is in fact false—that we can settle on a *particular* set of judgments to be regarded as erroneous. Which consistent set should be regarded as veridical? There are many candidates. But matters are yet worse than this, for the way in which we resolve the question which measurement judgments are veridical is to decide that they are *all* pretty likely to be false. We move on to Stage III and adopt a theory of errors of measurement (for more detail, see Kyburg 1984) according to which these errors are distributed approximately normally. But this means that when we judge that an experimental object, say, is the same length as a certain portion of a meterstick, we can be practically certain that our judgment is wrong.

Now this is really strange. Our progress toward objectivity, toward the sorts of sentences that might play the role of observation sentences in science, is marked by increasing uncertainty. We can resolve this oddity, as Mary Hesse (1974) does, by distinguishing between *observation reports* and *observation sentences*. The observation *reports* contain the results of our measurements (e.g., "The measurement of the length of the experimental object yields 0.876 meters"), and construed as judgments of length are wrong with a probability of one. The observation *statements* concern the objective magnitudes being measured (e.g., "The length of the experimental object is 0.876 plus or minus 0.002"), and are to be construed as being highly probable, relative to the observation reports. This makes very good sense, but it already introduces a layer of probability between our observations (observation reports) and the deductive consequences of our hypotheses and theories. It should also make us suspicious that there will be problems in accepting *conjunctions* of observation statements: Each of *n* observation statements may be rendered practically certain by a set of observation reports, but their conjunction may not be practically certain. In fact the *denial* of their conjunction may be practically certain, if *n* is relatively large. The deductive model of inquiry is being stretched out of recognition.

According to the deductive model, the set of statements representing our scientific knowledge is to be deductively closed. It is thus that we can justify the arguments that employ a number of premises to yield predictive or explanatory conclusions. Of course if the corpus is deductively closed, it must be consistent—else it would include all statements whatever. But then we must ask whether it includes statements that are merely highly probable? For example, the denial of Boyle's law in the example just given? Or even observation *statements*, as they have just been characterized? If not, it is void of interesting empirical content: its empirical content will be restricted to observation reports—the past incorrigible historical record. But then how can it serve to yield the premises required for prediction and explanation?

If our scientific corpus contains statements that are merely probable, we face two problems. First, what are these statements probable relative to? Not the body of scientific knowledge itself, for then they would not be merely probable but have a probability of one. And not the set of observation reports, for we require statistical generalizations as well as observation reports to render laws probable or improbable. Second, are uncertain statements incorporated into a body of knowledge merely because of their probability? If so, the lottery paradox is not far away. Or are there constraints that allow the corpus to be consistent? If so, what are they? There haven't been a lot of viable alternatives offered.

As a final possibility, suppose that at least some of the statements in our scientific corpus are accepted into that corpus on the basis of (probable) evidence, but are not themselves to be regarded as probable. If they are construed as having empirical content—as being possibly false—what *is* the warrant for accepting them? Why should we regard predictive and explanatory arguments employing them as premises *sound*? Is it merely a matter of convention and social goodwill—getting along with our peers? Or if they are not construed as having empirical content, how do they play their role in explanation and prediction? Again the deductive model is being distorted beyond recognition.

The Inductive Model

I have not offered anything that can pretend to be a fair characterization of the deductive model and its shortcomings, and I certainly haven't dealt with the various ways in which people have attempted to resolve some of the problems. But I hope that my sketch has suggested ways in which problems and puzzles seem to multiply. An attempt to solve one problem raises others, and a different attempt raises a whole new group of problems. Many of these problems have caused various writers to abandon various parts of the traditional deductive model; but often, it seems to me, the good and useful has been thrown out with the fantastic and foolish.

In the present section I shall sketch an alternative approach. It has much in common with the traditional deductive model, except that the basic and interesting relation is that of probability rather than that of logical consequence. This has the consequence that each of the core characteristics of the deductive model must be regarded as false:

(1) There need be no protocol vocabulary; no observational judgments need be regarded as infallible; no observation sentences need be regarded as incorrigible. This is so even in a given context or at a given historical moment.

(2) Scientific laws, theories, hypotheses—even universal generalizations—will be regarded as devoid of empirical content. They will be regarded as features of our scientific language—con-

ventional in the sense that language is conventional but *not* conventional in the sense that one convention should be regarded as "just as good as" another.

(3) The corpus of scientifically acceptable statements will be taken to be neither deductively closed nor fully consistent. Statements will be accepted on the basis of their probability or on the basis of the desirability of accepting them as a priori features of our scientific language.

(4) The only general statements that will be accepted on the basis of observational evidence will be statistical statements (the modal mob may substitute chance statements, if they insist), and most of these will concern errors of observation. Statistics and statistical inference will emerge as fundamental even in the most "deductive" parts of science.

Given these profound differences, how can I claim that the inductive model has much in common with the deductive model? That remains to be seen; but the idea is that where deductivism has a lot of plausibility, it can be seen as providing an *approximation* to the inductive model. The profound differences are important primarily in detail and in regard to foundational issues.

As in many variants of the traditional deductivist approach, we suppose that it is possible to provide a representation of (some fragment of) scientific knowledge in a formal language. We suppose that this language is rich enough to contain set theory and whatever else is required for a relatively sleek and efficient representation.[1] The notion of probability that I have defined in various places (e.g., Kyburg 1974) applies straightforwardly to such languages. Probabilities reflect formal logical relations and are represented by intervals. The domain of the probability function consists of ordered pairs, the first item of which is a set of sentences representing a body of knowledge, and the second a sentence: $P(K, S) = [p, q]$. The sets of sentences K are not required to be deductively closed, nor even consistent. We require only that if T is a member of K, and T and the axioms of the language entail R, then R also should be a member of K. (We shall say more about the axioms shortly; for the moment we assume that they at least contain axioms for set theory, and may also include "meaning postulates" characterizing the empirical predicates of the language—for example, postulates expressing the fact that "longer than" is an ordering relation.)

The sentences S fall into equivalence classes determined by K;

formally the equivalence relation is "is connected by a biconditional chain in K to." Every statement in the equivalence class of S receives the same probability as S. All probabilities arise from *direct* inference and are based on relative frequencies (or chances) known in K. The complicated problem (whose details—laid out in Kyburg 1983—I shall spare you) is that of providing criteria for the choice of a reference class, or, alternatively, providing an epistemic metalinguistic characterization of randomness. The definition of probability then goes roughly as follows.

$P(K, S) = [p, q]$ if and only if there are terms $a, b, c,$ and d of the language such that:

(1) a is a random "member" of the reference "set" c, with respect to K, and the sentence "$b(a) \, \varepsilon \, d$." (The scare quotes are intended to remind you that since the probability relation is being construed metalinguistically, the relation of randomness concerns terms and sentences rather than objects, sets, and states of affairs.)

(2) b is a quantitative function (a random variable, denoting a random quantity).[2]

(3) d is a term denoting a Borel set, not containing definite descriptions or the like.

(4) "$b(a) \, \varepsilon \, d$" is in the equivalence class of S.

(5) K contains the statement that the proportion of items in c that have b-values in d (or the chance of c items' having those values) lies between p and q.

It should be noted that there may be many such quadruples a, b, c, d; in fact, given the resources of set theory we can easily prove that there are an infinite number. But of course they all yield the same probability interval.

There is a slight complication in that not every term definable in the language can be allowed to serve as a potential reference term in the language, and not every functor in the language can be allowed to serve as a random variable. Since we will in due course consider the replacement of one language by another, we will suppose here that the language L includes a recursive specification of the reference terms C and the random variables B, in addition to the specification of its predicates and operators. We shall also include as part of the language a recursive specification of its axioms, which will include definitional axioms, axioms for set theory, meaning postulates, and so on.

Since we are looking at things probabilistically, we had best

index our bodies of scientific knowledge according to the probabilities of the statements we have in them. Simultaneously, probabilities are relative to those bodies of knowledge. In some contexts
we want to accept only statements of which we are very sure
indeed; in others our epistemological demands may be less severe.
I shall consider that the corpus of level r, K_r, consists exactly of
those statements whose lower bound of probability, relative to a
corpus of higher level, is at least r. This has the immediate consequence that a corpus will be neither deductively closed nor (in most
ordinary cases) consistent. Whether this is something to be upset
about or not remains to be seen. For the moment it suffices to
observe that probabilities *can* be defined relative to such corpora.

It should be noted that approaching bodies of knowledge in this
way amounts to the introduction of two sorts of uncertainty. Both
are related to probability. One just *is* probability: we speak of the
probability of a statement S relative to a body of knowledge. But
that body of knowledge itself contains uncertain statements—statements that are merely probable relative to a corpus of higher level.
For the purpose of evaluating the probability of S, those statements,
though in a sense uncertain, are treated as certain: they are treated
as *evidence* for the evaluation of the probability of S. Implicit in
the statement that the probability of S is $[p, q]$ is the standard of evidence according to which probabilities of less than $1 - r$, relative to
a higher level corpus, are to be regarded as practical impossibilities.

Statements can get into a body of knowledge of level r by having
a probability of at least r, relative to a higher corpus of knowledge.
But they can also get into a body of knowledge of level r as a result
of observation. Note that there is nothing in this that requires us
to suppose that observation is error free.

The procedure of accepting uncertain observation statements
directly into the corpus K_r can be given a rational reconstruction
along the same lines as that given above for the acceptance of statements into a corpus K_r in virtue of their probability relative to a
higher-level corpus. In this case it is the probability of *error* relative
to a *meta*corpus, that determines the level of acceptance of the
observation statement.

Suppose that A and B are "observation predicates," and that an
individual has observed a lot of entities that are both A and B; a
number of entities that are A but about which he has not had an
opportunity to judge whether or not they are B; and a small number

of entities that are A and not-B. Were he able to accept the generalization "All A's are B's," he could, in the second, as-yet-unjudged cases, also accept the *prediction* that the entities involved are also B's. For simplicity, let us suppose that his judgments about A's are infallible. And let us suppose that he *does* accept the generalization that all A's are B's. I shall return to the grounds on which universal generalizations, laws, and theories can be accepted in the next section; we shall leave that to one side for the moment. It follows from his acceptance of that generalization, that some of his "observations" of non-B's have been erroneous. It is a mere matter of metalinguistic bookkeeping for him to note that of N acceptances of statements of the form "not-$B(x)$" that he has made, R must be in error—given his acceptance of the universal generalization. This set can be made the basis of a statistical inference to the general reliability of "not-B" judgments. We require that it be a random sample of such judgments, in the appropriate epistemological sense, and that is essentially *all* that is required for him to be able to accept, in a metacorpus of level r, MK_r, the statistical generalization that between ε_1 and ε_2 of such judgments are in error.

Relative to the metacorpus of level r, then, the probability that a *random* judgment of the form "not-$B(x)$" is in error is the interval $[\varepsilon_1, \varepsilon_2]$. The probability that it is *not* in error is thus $[1 - \varepsilon_2, 1 - \varepsilon_1]$. Such a judgment is acceptable in the corpus of level $1 - \varepsilon_2$, and of course in any corpus of lower level.[3] A particular judgment, say "not-$B(a)$," may or may not be random in the required sense: if "$A(a)$" is in the corpus, and experience impels us toward the acceptance of "not-$B(a)$," we *know* that the tempting judgment would be in error. (Remember that we have supposed the application of the predicate A to be error free, and that we have taken the generalization "All A's are B's" to be accepted.)

There are a number of things to be noted about this simplified special case. First of all, note that we do not collect statistics on error by counting first veridical and then erroneous judgments of the form "not-$B(x)$." If we can suppose that we can make errors in such judgments, there is no way we can be sure that a given judgment is veridical.

Second, we can in this particular case know that a certain judgment is in error: if we have "$A(a)$" in our corpus, and judge "not-$B(a)$," we know the latter judgment is in error. But this is not typical. We generally will not have infallible judgments at our service: "$A(a)$" may be the judgment that is in error. And in instances involving a number of judgments ("is the same length as" provides an example) all we may know is that *somewhere* in the list of statements supported by observation there are at least R that are wrong. There is no reason to suppose (although we did so for the sake of the example) that *any* empirical predicates in the object language can be applied without risk of error.

Third, there is no purely logical reason for supposing that only those statements that conflict with other statements in our corpus are wrong—we *might* suppose that all our "not-B" judgments are in error. Perhaps everything is B. But such an assumption is gratuitous, it flies in the face of our desire to maximize information, and it is probably linguistically unintelligible. We therefore assume minimum rates of error compatible with our bodies of knowledge at a given time, and project those rates, via statistical inference, to future bodies of knowledge.

Fourth, this treatment of observational error can already introduce inconsistency into the corpus K_r. Under *perfectly ordinary circumstances* a person can reasonably judge, and accept into his corpus, the assertion that a is not longer than b. But it is clear that there are circumstances under which each of a series of judgments of the form "a is not longer than b" may be acceptable, while the set of these judgments as a whole conflicts with the axioms of order. It might be suggested that, under these circumstances, the rational agent should suspend judgment about all of the observation judgments in the set—at least when there is no more reason to be suspicious of one than of another. It seems to me that this would hamstring science. We must be able to accept uncertain observation statements, and since they are uncertain, we can have no guarantee that they will not be mutually inconsistent. It might alternatively be suggested that what is reasonably acceptable is some consistent subset of the full set of these judgments. But the arbitrariness of this is offensive; the doxastic states of rational men under identical circumstances should be the same; and we would lose the power to discriminate between a set of rational beliefs all of which are required and a set of rational beliefs that is an arbitrary subset of those permitted.

Let us now consider the question of empirical content. Clearly the (uncertain) observation statements that are accepted into a corpus of level r embody empirical content, whether they are accepted on the basis of observation, on the basis of statistical inference, or through deduction from already accepted statements. Furthermore, according to the conception of probability we are employing, statistical statements relating observable predicates (those predicates that appear in acceptable observation statements) can come to be accepted: we can accept into a corpus of level r—on the basis of observations that appear in a higher-level corpus—statements of the form "The proportion of X's that are Y's lies between p and q." We may even accept general distribution statements: "The quantity B is (approximately) distributed in the set A according to a distribution in the family D." But since our statistical data are always limited, we can never render probable a statement of the form "The proportion of X's that are Y's is exactly p," and therefore we can never, on the basis of statistical evidence, accept a statement of the form "All X's are Y's." I shall argue shortly that such statements—universal generalizations, laws, and theories—are without empirical content.

But there is one further sort of statement that embodies empirical content: the statements in our metacorpus that describe the frequency with which we erroneously accept statements of a given kind on the basis of observation. Our knowledge about errors of observation is empirical knowledge. To say that in distinguishing between C's and non-C's we are wrong between p percent and q percent of the time is to say something substantive about our observations of the world.

In the case of measurement—which is of much more interest to us than mere classification, since laws and theories are of more interest to us than universal generalizations—we can establish general statistical inferences that give us knowledge not only about the frequency of error but also about the frequencies of errors of different magnitudes. These statements can be rendered probable and acceptable, in a metacorpus MK_r, on the basis of a higher-level corpus containing data on the required frequency of rejection of various statements among those that we have considered up to a given time.

As a specific example, consider (yet again) the transitivity of the relation "is not longer than." Ordinarily a reasonably large set of judgments of the form "a is not longer than b" will be inconsistent with the generalization articulating this transitivity. We have two ways of approaching consistency: to suppose that the generalization is false, since it is refuted by observation—in which case we will have to do without quantitative science—or to suppose that the generalization is accepted a priori and serves as the *standard of error* in the set of particular judgments. But that means that the generalization is irrefutable, and therefore, *in a certain sense,* without empirical content. I say "in a certain sense," for, as we shall see, it is possible to give criteria for the preferability of one language, containing one set of a priori conventions, to another.

If the theory of measurement can be taken as paradigmatic, then we shall be led to regard all quantitative laws as forming an a priori, irrefutable net. To this I would add ordinary universal generalizations: what are represented as universal generalizations ("All crows are black," for example) either admit of exceptions, and are thus really better construed as statistical generalizations that bear parameters close to zero or close to one, or are best construed as a priori standards of observational error.

Choosing among Languages

It is time to consider the grounds for the choice among scientific languages. This is clearly crucial for the program considered here. But it should be noted that it also presents a problem for the deductivist. We know that scientific progress is often accompanied by the introduction of new terminology ("quark," "charm," "oxygen," "atomic number") as well as by changes in old terminology (reclassification in taxonomy, classical as opposed to relativistic "mass"). These changes are not easy to account for on the deductive model. Talk about research programs, scientific revolutions, and the like may in fact be construed as motivated precisely by these difficulties. They are clearly something that must be grappled with.

On the view defended here, however, the problem of linguistic change in science is even more pervasive. On this view scientific laws and generalizations themselves constitute part of the *meanings* of scientific terms; to change a law, to adopt or reject a generalization, even to refine a generalization, on this view, is to change the language of science. Since these laws and generalizations are taken as a priori, they are to be construed as partially constraining the meanings of the terms that occur in them. Scientific language must be viewed as being in a constant state of flux.

Nevertheless, we are abandoning the deductive model, so incommensurability (that is, logical incommensurability) need not bother us. Since generalizations are not refuted by counterinstances anyway, we need not struggle to show that a generalization at one stage of science using one language can be inconsistent with a generalization at another stage using a different language. New observations expressed in one language need not be regarded as "refuting" old generalizations expressed in another language.

Consider a language L consisting of a set of predicates (P), operators (O), random variables (B), reference terms (C), and a set of statements A taken to be a priori and characteristic of the language: $L = <P, O, B, C, A>$. Given a certain body of experience, rational corpora of various probability levels are generated for this language. A statement enters a corpus of a given level provided its probability (or 1 – its probability of error), relative to a corpus of higher level, exceeds the given level. Thus a corpus of a given level may contain statements that are warranted directly on the basis of observation ("The voltmeter reading is '10.2 volts'"); statements that are unlikely to be in error ("The voltage is between 10.0 and 10.4 volts"); statistical statements ("In 40 to 50 percent of the trials, in the long run, the voltage is less than 20 volts"); and, in a metacorpus, error statements ("The difference between the voltage reading and the true voltage, in measurements of kind V, is approximately normally distributed with a mean of O and a standard deviation of .1").

Furthermore, these corpora will contain predictive observational content: if the generalization "All crows are black" is part of our language, and if "a is a crow" is in our corpus, "a is black" will also be in our corpus, whether or not it is there as a result of observation. If, according to our metacorpus, "black" is a term whose application is relatively error free, we will be justified in regarding "a is black" as a statement that *could* be justified by observation.

It is these statements that could become warranted by observation that embody the *useful* empirical content of a rational corpus. Given a metacorpus containing knowledge of errors of application of various terms in the language, we can characterize a set of sentences that in principle could come to be accepted directly in a corpus of level r on the basis of observation. They are basic (noncompound) sentences employing terms whose observational application is relatively error free (i.e., such that the chance of error in applying them is less than $1 - r$).

A number of these statements will occur in the corpus of level *r.* Some of them will be there as a result of observation; some on the basis of direct statistical inference; some on the basis of deductive inference. (Of course, some may be included in the corpus on more than one basis.) From the set of observationally warrantable statements in the corpus of level *r*, delete those that are in fact warranted by observation. What remains is a set of basic sentences warranted by deductive inference or by direct statistical inference. We shall call this the *novel observational kernel* (*NOK*) of the corpus of level *r*. It is unlikely that this set is finite, in the case of any reasonably strong language and any reasonably broad experiential base. Let us suppose that we can render it finite through the application of some general spatiotemporal constraint *C*. The result we denote by NOK_C.

The content of NOK_C depends on the level of rational corpus we have chosen to look at; it depends on the language (including its a priori elements) that we have chosen to employ; and it depends on the body of experience we are taking account of, both through the acceptance of statistical generalizations in the rational corpora of levels higher than *r* and through the acceptance of statistical generalizations concerning error in the metacorpora. We can express these dependencies by writing $NOK_C(r, L, E)$.

Given a body of experience *E* (we need have no way of *describing* this experience, or representing it linguistically—*E*, of course, is not a set of sentences), a level of corpus *r*, and a constraint *C*, we can compare two languages, L_1 and L_2, by looking at the cardinalities of $NOK_C(r, L_1, E)$ and $NOK_C(r, L_2, E)$. The language with the larger novel observational kernel is preferable.[4]

This is all programmatic, and in a variety of senses problematic. We have not characterized *E* at all, and we know that what we experience, even on a prelinguistic level, depends in some degree on what we understand and believe, as well as on the language we speak. Furthermore, no two individuals have exactly the same body of experience. Nevertheless, it may be conjectured that the effective differences may be small enough to be disregarded, at least with respect to the parts of experience relevant to scientific inquiry. Even if this is not true in general, it is certainly very nearly true when it comes to a society of scientists working in a given area.

It is also a matter of conjecture that the level *r* of a corpus chosen for analysis will not make a crucial difference—that if *C* and *E* are held constant, we will not find L_1 preferable to L_2 if we look at level *r* and the opposite to be the case if we look at level *r'*.

Nor have we spelled out the constraints *C*. We merely hope that two plausible constraints *C* and *C'* will not lead to opposite results.

Nevertheless, as we shall see in the next section, it is possible in certain special cases to apply the criterion of the larger *NOK*, and in fact often to justify the deductivist picture as a reasonable approximation to the truth in certain special cases.

The Justification of Deduction

We are now in a position to relate the deductive and inductive models of scientific inference, and to show why the deductivist is often right to a first approximation. Clearly there are lots of cases in which the deductive model seems to represent scientific inference fairly well.

Perhaps the most obvious is the application of scientific theory in making predictions, in designing machinery to achieve certain ends, in providing explanations, and the like. The general scheme might be presented like this: We have a general law or theory *T*. We note (if we are making a prediction or providing an explanation) or create (if we are building a better mousetrap) a certain set of boundary conditions represented by the observation statement *B*. From *B* and *T*, a sentence *O* follows deductively, where *O* is the prediction or the outcome we wish to achieve or explain. (Obviously this is only one pattern of "explanation" among many.)

On the deductivist view, there are three problems: the grounds for accepting *T*, the grounds for accepting *B*, and the fact that if *O* doesn't turn out to be quite as anticipated, we do not generally regard this as refuting *T*. These are problems, because if we lack grounds or justification for accepting *T* or *B*, the argument is pointless, however valid it may be. If the argument is to be more than a way of bullying other people, it must be epistemically *sound* as well as valid.

On the view presented here, *T* is regarded as a priori, and therefore eminently capable of functioning as a premise in an argument. *B* is regarded as acceptable but not incorrigible: it admits of the possibility of error. Nevertheless, it is probable enough to be accepted into a rational corpus, and thus yields, in conjunction with the a priori *T*, the deduction of *O* in that corpus. Sentence *O*, again, is an observation statement, but not one that need be regarded as incorrigible. We obtain *O* in our corpus through deduction; if subsequent observation fails to yield observational warrant for *O* (or provides warrant for $\ulcorner \sim O \urcorner$), this constitutes an item of data that is added to the basis of our statistical theory of error. Either O or B (or both) will be thought of as having been "corrected." Note that no prediction is perfect, and that no machine ever functions exactly as it was designed to function. One of the main jobs of a research and development group is to figure out why a new design is not working as predicted. When we are all through—when it does work the way it is supposed to—then the deductive model applies beautifully. Parents are all deductivists when they explain locomotives to their offspring.

In reality we often generate predictions out of a number of laws and theories, combined with a large number of statements representing boundary conditions. The deductivist view imposes deductive closure on our bodies of knowledge, so that this case is easily reducible to the previous one. On the present view, this reduction is not automatic, since we have not imposed deductive closure on our bodies of knowledge. On the theoretical level, since laws, theories, and generalizations are all regarded as a priori, we do have deductive closure; we may combine the theoretical premises for our deduction into a single premise T. But the boundary conditions for our deduction may involve a number of observationally warranted statements. What is required for the deduction to lead to the inclusion of O in the rational corpus is that the *conjunction* of these boundary condition statements be acceptable in that corpus. The acceptability of this conjunction will of course depend on the distribution of errors of application of the terms that appear in them, but in general the more boundary condition statements are required, the larger the error in their conjunction. (There are exceptions: the variance of the error of an average of observations will be less than the variance of the error of a single observation.) A deduction from a large number of observational premises will thus be more insecure—will be capable of being carried out only in a relatively lower rational corpus—than a deduction from a smaller number of observational premises. This phenomenon is reflected directly in the inductive model.

Deductive argument is thus relevant and useful in the inductive model, but is subject to constraints and limitations that are hard to express in deductivist terms. Furthermore, where the status of the premises and conclusions is somewhat obscure in the deductive model, their status can be rendered quite clear in the alternative model.

Let us turn to the classical examples of falsification and verification that are almost always cited by deductivists—even reformed deductivists—as though they were paradigmatic. Consider the (so-called) empirical generalization "All A's are B's." We are supposed to test this generalization by obtaining an A and examining it to see whether or not it is a B. If it is not a B, we have in one fell swoop refuted the generalization; if it is a B we have confirmed, in some slight degree, the generalization.

On the present view, universal generalizations (not to mention laws and theories) are regarded as a priori. Thus they cannot be refuted, and have no need of confirmation. But something like the situation described classically does seem to occur. How can it be represented in the inductive model, and, more important, what are the circumstances under which it is an approximation to the truth?

Note first that neither "a is an A" nor "a is a B" nor the denials of these statements are to be regarded as incorrigible. They are (we suppose) observation statements, and as such they are subject to error. Consider two languages, in which "All A's are B's" is an a priori component, and L_2, which lacks this component.

In general we will have a large number of observations of A's, of B's, of not-A's, and of not-B's in our rational corpus. Let's suppose, further, that in our language we have a number of (a priori) generalizations connecting each of these four predicates with other observational terms. It is thus, and only thus, that we have the data that will yield error frequencies able to serve as the basis for accepting into our metacorpus generalizations concerning the frequency of error in the application of each of the four predicates B, A, not-B, and not-A.

Now consider the choice between L_1 and L_2. According to the deductivist, a single observation of an A that is not a B should settle the matter. In view of the fact that we cannot apply these predicates with complete security, I think this should never happen. But it is possible, if we very rarely make mistakes in applying the predicates A and not-B, that a few "negative" observations will suffice to determine the choice between L_1 and L_2, particularly if those observations belong to a class of observations in which error is extremely rare—that is, they are made carefully, by a number of people, in good light, etc. The mechanism is this: If we speak L_1, we must, in the face of these observations, suppose that the attribution of the predicates A and not-B on the basis of observation is more perilous than we thought. Now this not only affects the usefulness

of the generalization that all A's are B's but also affects the usefulness (in generating predictions) of *other* generalizations that use the
predicates A, B, not-A, and not-B. Thus there comes to be a very
rapid loss of items from the NOK of our rational corpus.

There is an additional factor that can play an important role in
bringing this about. We supposed that there were a number of
generalizations, other than the one at issue, which employed the
predicates in question. In many such cases it will have been possible
to provide special techniques of observation that yield error rates
very near to zero, or that at any rate result in improved dependabil-
ity in the application of the predicates. If that is so, and the recalcitrant observation does not yield to these special techniques of
observation, then to adopt L_1 is to forgo even more of the benefits
of the other generalizatioins involving the predicates in question,
because doubt will also be cast on the efficacy of the special observational techniques developed in response to them.

In short, falsification by a counterinstance occurs when the application of the terms involved is highly dependable, and the assumption that the alleged counterinstance is merely an artifact of observational error is going to require a significant change in our assessment of frequencies of error, with the result that we will lose more
items from our NOK than we would gain by having the generalization in question form a part of our language.

How about the "slight confirmation" alleged to be provided by
positive counterinstances? We have here (other things being equal)
an observational attribution of A and an observational attribution
of B that we have no reason to reject whether we speak L_1 or L_2.
These observations constitute a contribution to the evidential base
of the statistical generalizations concerning the frequencies of error
in the use of these terms, and contribute to our assessment of their
application. A number of such observations may have a significant
impact on our "theory of error" concerning them—particularly
when the evidential base is otherwise not large. The preferability
of L_1 over L_2 is increased not only because, in the presence of the
generalization that all A's are B's, the NOK of a corpus based on
L_1 is larger than that of a corpus based on L_2 but because an
increase in the (statistical) grounds for thinking that A and B can
be dependably applied may increase the general usefulness of those
terms, and add to the NOK in ways that have nothing to do with
the generalization in question. It is nevertheless clear that the impact
of the "positive instance" must generally be regarded as very slight.

It is doubtful that much of interest in science concerns qualitative universal generalizations. Quantitative laws and theories are of far more interest, and here the connections and contrasts between the inductive and deductive models are far more enlightening. Consider, as a simple example, the Boyle-Charles gas law: $PV/T = P'V'/T'$ for a given quantity of gas. According to the simple deductivist account, the law should be refuted by a sextuple of observations on a sample of gas, if those observations don't satisfy the law. But this is ridiculous: without in the least calling the law into question, we *expect* sextuples of observation *reports* to fail to satisfy the relation. Any lab assistant would be justifiably suspicious of a freshman who reported a sextuple of observations for which $PV/T = P'V'/T'$ held.

We may distinguish observation *statements* from observation *reports*. An observation report is a sentence that reports the experience of the student. The observation statement is the "observable" fact of the matter. Thus if the Boyle-Charles law is true, there can be no sextuple of observation statements failing to satisfy the law. What is the relation between observation reports and observation statements? According to the usual treatment of errors of measurement, errors of any magnitude are *possible*, though big errors are rarer than small ones. *Deductively*, if we take account of the possibility of unlimited error, no sextuple of observation reports is ruled out by the truth of the law when we take the law to concern observation statements.

But we do not ordinarily suppose that arbitrarily large errors need to be taken account of. Given an observation report, together with a known distribution of error for that *kind* of report, we infer—not deductively, but inductively or probabilistically—the observation statement that constrains the quantity in question to lie within a certain interval. Thus, if our observation report is that a temperature is 373.15 degrees, and the errors of measurement of temperature corresponding to the technique of measurement we use are distributed approximately normally with a mean of 0.0 and a standard deviation of .05, the probability is .99 that the temperature lies between 373.02 and 373.28 degrees—*provided* that this measurement is an epistemologically random one with respect to embodying an error less than .13. Thus the sextuple of observation statements acceptable in a corpus of level .99 would have the form:

P lies between p_1 and p_2.
V lies between v_1 and v_2.

T lies between t_1 and t_2.
P' lies between p'_1 and p'_2.
V' lies between v'_1 and v'_2.
T' lies between t'_1 and t'_2.

But while each of these statements may have a probability of .99, what we need for the refutation of the Boyle-Charles law is not that set of statements but their conjunction; and obviously the probability of their conjunction is much less than .99 (in fact if they are independent—and there is no reason we should not regard them as independent, unless we are prepared to accept the law as a priori—the probability of the conjunction would be on the order of .94).

With the help of a natural convention or two, we can come up with a six-dimensional region within which our pressures, volumes, and temperatures fall with an epistemological probability of .99: we can accept in our corpus of level .99 that this region includes the true values—provided, of course, that the conditions of randomness are met. If there is no sextuple in this region that satisfies the relation $PV/T = P'V'/T'$, we might be able to say (in statistical parlance) that the Boyle-Charles law was rejected at the confidence level of .99.

This is some distance from the deductivist account already; but we're not through yet. We must inquire into the conditions of randomness, and into the source of our knowledge of the distributions of errors of measurement. Obtaining distributions of errors of measurement is a complicated business, which I cannot go into here. But it turns out that in order to have quantitative criteria of error, we must already have accepted quantitative laws. It is against the background knowledge of such laws that we obtain our knowledge of the statistical distribution of errors of measurement. If this is so, then it may be that the Boyle-Charles law itself plays this role, and thus that what the sextuple of observation reports provides us with is a new item of data regarding *errors of measurement* rather than data bearing on *the acceptability of the law*. This would surely be the case if, for example, we were measuring temperature by a gas thermometer! This is easily and naturally reflected in the inductive reconstruction: if the law is taken as part of the background knowledge, then a set of reports that generated a region in which the law was not satisfied would not be epistemologically random in the required sense. Observations of pressures, temperatures, and volumes *could* not lead to a rejection of the law; they could only lead to new data on which to base our knowledge of the distribution of errors of measurement of these quantities.

But isn't it possible for observation to lead to the rejection of the Boyle-Charles law? Didn't that in fact happen historically? This is no doubt true in some sense, but the law was not rejected in response to a contrary instance. In the first place, the law can be used as part of the framework relative to which we develop and refine our theories of errors of measurement; there is an interplay between the law and the laws underlying the measurement of the quantities involved—in this case, particularly the measurement of temperature. One criterion of improvement in our techniques of measurement is that they lead to results fitting the law. At the same time, one discovers limitations of the law: the same improvement of techniques that leads to improved agreement between measurements and the law in the case of hydrogen at high temperatures leads to increased disagreement in the case of carbon dioxide at high pressures. The scope of application of the law becomes narrower; but for certain instances—hydrogen, for example—conformity to the law is still a *criterion* of accuracy of measurement.

Furthermore, we can sort gases out according to the degree to which they conform to the law: the law gives poorer results for heavier gases or gases at temperatures and pressures near identifiable changes of state. It is natural then to construe it as a law concerning *ideal* gases—precisely because we have independent criteria for degrees of ideality. This amounts, finally, to taking the law to have an empty scope—to be true of no real gas. But this does not amount to *rejecting* the law; it amounts to taking it as true a priori for ideal gases (of which there are none), which is a useful thing to do precisely because we have ways of *measuring* the departure of actual gases from ideality.

Of course we can *imagine* a different history—one in which the Boyle-Charles law (that is, from the present point of view, the language embodying the Boyle-Charles law) comes to be flatly rejected. We could suppose that the result of speaking the Boyle-Charles language was disorganization in our knowledge about errors of measurement. Our observation reports concerning samples of gases, for example, might oblige us (in that language) to suppose that errors of measurement of temperature are more widely dispersed than we had hitherto thought. This would have the result of rendering unacceptable temperature predictions that we had previously been able to accept. This could lead to a rapid loss of content in our bodies of knowledge; the deductive model reflects this situation to a first approximation.

In order to find instances of falsification that might conform better to the deductivist account, we must turn to crucial experiments that decide between very general theories. Everybody admits that this happens but rarely. One case that is often cited is the Michelson-Morley experiment. The first thing to be noted about such experiments is that they are complex and presuppose a vast amount of physical theory. By "presuppose" I do not mean that in some mystical way these laws are "used" in designing apparatus and in determining results without any serious epistemic commitment to them on the part of the designers. We must have *reason to believe* that the experiment works the way it is alleged to work, that the apparatus will function as required, and so on. Note that in the characterization of the apparatus, we call not only upon certain laws and generalizations, which I would take as general a priori features of both alternative languages, but also upon physical constants and measurements, which can only be known in a probabilistic sense, and then only as lying within certain limits. Furthermore, we need the *conjunction* of all this material in a rational corpus in order to obtain the predictions of the two theories.

The second thing to be noted is that the observations themselves generally involve measurement, and thus are probabilistic in character. Given Theory I, and given our corpus of practical certainties—that is, the corpus of statements that in the context may be regarded as "practically certain"—we can infer that a certain quantity will have a value in a certain range; and given that the quantity has a value in that range, we can be practically certain that our measurement of that quantity will fall in another, somewhat larger, range. Similarly for Theory II.

The third thing to be noted is that as a matter of historical fact the results of such crucial experiments as the Michelson-Morley are rarely as definitive as they are painted to be in the textbooks. Not only are they not definitive logically (we have already seen that) but they are not even regarded, immediately, as definitive by experts in the field. This hardly conforms to a deductivist account, since a deduction is either valid or invalid, and people can recognize which rather rapidly, and *that's* surely not what the old guard is hesitating about.

On the inductive model, what we are choosing between are *languages*, and our criterion of choice involves the *NOK* of a rational corpus of practical certainties. The choice between two such rarified theories as those at issue in the Michelson-Morley experiment doesn't seem to conform to this model very directly, either. Can we really claim that relativistic mechanics, combined with the results of the Michelson-Morley experiment, leads to a body of practical certainties with a larger *NOK*? It seems rather unlikely—particularly at the time of the experiment. The predictive advantages of relativistic mechanics over Newtonian mechanics were, at the time, surely negligible.

This seems to me a case in which neither theory—that is, neither language—offers immediately demonstrable advantages over the other. Yet we must, faced with the choice, make a choice, if only sooner or later. Given two theories that are strikingly similar as far as ordinary predictions are concerned, we naturally turn to the consideration of hypothetical predictions: suppose our measurements could be made more precise (as no doubt they will be); suppose we could take account of more details in our deductive analysis of the experimental setup (in principle, we probably always can); suppose we could give a *quantitative* account of the effect of the simplifying assumptions we have made; and so on. These are things we have often been able to do; they represent not "mere" epistemic possibilities but realistic ones. And it is often possible then to *show* that under these hypothetical circumstances one language will yield a more informative body of knowledge, when combined with a natural and plausible extension of our experience, than the other.

It is this *sort* of thing that I think Ronald Laymon has in mind when he speaks of the "modal auxiliary" associated with a scientific argument (1980), or when he writes:

> Newton's argument against the received view is not that his theory fits the facts better, but that the received view can do no better than it now does. This argument suggests that experimental fit *per se* . . . is not as important for theory testing as is the demonstration that an existing fit can or cannot be improved. (1978, pp. 249–250)

There is thus plenty of room for deduction in the inductive account of scientific explanation and inference, and on occasion the deductivist account even yields approximately the right answers.

There are two ineradicable differences. First, on the inductive account there is *always*, in every case, a layer of probability interpolated between knowledge and experience. No hypothesis, theory, law, or even observation statement ever entails, or logically excludes, any observational judgment. Second, on the deductivist account premises may be piled on premises indefinitely without weakening the force of a deductive argument, while on the inductive account the soundness of a deductive argument hinges in part on the acceptability of the *conjunction* of its premises, which often decreases with their number. If we can accept uncertainties we had best not demand consistency or deductive closure. These differences entail constraints on the role of deduction, but they also provide a justification of deduction within those constraints.

Notes

1. Being thus powerful, sets of sentences in these languages will admit of unintended models. If this is bothersome, we may in our model theory restrict our attention to "intended" models; we should have no more difficulty doing this than those who prefer talk of models have in specifying the models thay take to be relevant to scientific theories.

2. In contrast to the usage of most statisticians, I take "random variable" to denote a linguistic entity—a functor—and the term "random quantity" to denote a set-theoretical entity—a function from individuals or sets of individuals to real numbers, magnitudes, or vectors of them.

3. I have slurred over an important distinction here between having grounds for taking a judgment "not-$B(a)$" to be in error and supposing that, lacking these grounds, we may take the judgment to be veridical. This requires argument, for which this is not the place. The argument depends on the pragmatic supposition that if I (or the community) *never* have to retract the statement "not-$B(A)$," then it is true enough for all practical purposes.

4. Note that what is at issue in computing these cardinalities is not the set of statements that *will* come to be admitted on observational grounds as well as on inferential grounds but the set of statements that are admitted on inferential grounds and *might* also come to be admitted on observational grounds. Various refinements and improvements suggest themselves. If, on one hand, our corpus contains the nonbasic observational sentence "If B then O," where B is a statement whose truth we could *bring about* (e.g., "I look at the patch of sky with coordinates beta, gamma at 0:15:36 on such and such a date"), then O should perhaps be counted as part of the

potential observational content of our corpus. On the other hand if a predictive observational statement *O* can be obtained from our corpus only by an enormously long and complicated computation, it should perhaps not count as much as a statement *O'* that can be obtained more easily. And perhaps we should even find some way of taking account of the relative importance to us of different sorts of statements.

References

Carnap, Rudolf. 1963. "The Aim of Inductive Logic." In E. Nagel, P. Suppes, and A. Tarski, eds., *Logic, Methodology, and Philosophy of Science,* pp. 303–318. Stanford, Calif.: Stanford University Press.

Feyerabend, Paul. 1970. "Against Method: Outline of an Anarchistic Theory of Knowledge." In M. Radner and S. Winokur, eds., *Minnesota Studies in the Philosophy of Science,* vol. 4, pp. 17–130. Minneapolis: University of Minnesota Press.

Glymour, Clark. 1980. *Theory and Evidence.* Princeton, N.J.: Princeton University Press.

Hempel, Carl G. 1965. *Aspects of Scientific Explanation and Other Essays in the Philosophy of Science.* New York: The Free Press.

———. 1970. "On the 'Standard Conception' of Scientific Theories." In M. Radner and S. Winokur, eds., *Minnesota Studies in the Philosophy of Science,* vol. 4, pp. 142–163. Minneapolis: University of Minnesota Press.

Hesse, Mary. 1974. *The Structure of Scientific Inference.* Berkeley, Los Angeles, London: University of California Press.

Hintikka, Jaakko. 1966. "A Two-Dimensional Continuum of Inductive Methods." In J. Hintikka and P. Suppes, eds., *Aspects of Inductive Logic,* pp. 113–132. Amsterdam: North-Holland.

Kuhn, Thomas. 1962. *The Structure of Scientific Revolution.* Chicago: University of Chicago Press.

———. 1977. *The Essential Tension.* Chicago: University of Chicago Press.

Kyburg, Henry E., Jr. 1974. *The Logical Foundations of Statistical Inferences.* Dordrecht: Reidel.

———. 1979. "Direct Measurement," *American Philosophical Quarterly* 16:259–272.

———. 1983. "The Reference Class," *Philosophy of Science* 50:374–397.

———. 1984. *Theory and Measurement.* Cambridge: Cambridge University Press.

Lakatos, Imre. 1970. "Falsification and the Methodology of Scientific Research Programmes." In I. Lakatos and A. Musgrave, eds., *Criticism and the Growth of Knowledge,* pp. 91–195. Cambridge: Cambridge University Press.

Laymond, Ronald. 1978. "Newton's Advertized Precision and His Refuta-
    tation of the Received Laws of Refraction." In P. Machamer and R.
    Turnbull, eds., *Studies in Perception,* pp. 231—258. Columbus: Ohio
    State University Press.
————. 1980. "Idealization, Explanation, and Confirmation." In P. As-
    quith and R. Giere, eds., *PSA 1980.* East Lansing, Mich.: Philosophy of
    Science Association.
Levi, Isaac. 1981. "Direct Inference and Confirmational Conditionaliza-
    tion," *Philosophy of Science* 48:532–552.
Nagel, Ernest. 1961. *The Structure of Science.* New York: Harcourt, Brace,
    and World.
Popper, Karl. 1959. *The Logic of Scientific Discovery.* London: Hutch-
    inson.

5. Deductivism Visited and Revisited

Wesley C. Salmon

Deductive chauvinism, as I remarked in the introduction to this volume, comes in two forms—inferential and explanatory. Inferential chauvinism is associated mainly with Karl R. Popper and his followers; surprisingly perhaps, it was also espoused—though in a rather different form—by Bertrand Russell.[1] My main argument against Popper's inferential deductivism is contained in "Rational Prediction," which is reprinted above.

In this paper I shall focus on explanatory deductive chauvinism. I shall pay it a brief visit in the context of Laplacian determinism, and then revisit it in the context of modern science, where there is a strong presumption that indeterminism holds sway. In the indeterministic setting, I shall argue, explanatory deductive chauvinism cannot prevail.

Deductivism and Determinism

Explanatory deductivism flourishes in the fertile soil of Laplacian determinism. Consider Laplace's demon—the imaginary being that knows all of the laws of nature and the precise state S_1 of the universe at just one moment, and is capable of solving any mathematical problem that is in principle amenable to solution. This being can provide a deductive-nomological (D-N) explanation of any particular event whatever. If the event to be explained comes after the special momentary state S_1, known in complete detail by the demon, then the demon can summon laws that, in conjunction with facts drawn from S_1, entail that the explanandum event occurs. This deduction constitutes a D-N explanation. If the event to be explained precedes S_1, the demon can make a retrodictive inference to facts preceding the explanandum event. These preceding facts constitute initial conditions that, in conjunction with the pertinent laws, entail the occurrence of the event to be explained. The statements describing the preceding facts, in conjunction with applicable law statements, constitute the explanans for an acceptable D-N explanation. The same general strategy will work if the demon wants to explain some event contained within S_1. Earlier facts can be inferred, and these can be used to provide the desired explanations. Moreover, the demon can construct a D-N explanation of any regularity, provided it is not one of the basic regularities that constitute fundamental laws of nature. These most general regularities—for example, for Laplace, Newton's laws of motion—have enormous explanatory power, but they cannot be explained, because there are no laws of still greater generality under which they can be subsumed.

Laplace realized, of course, that human beings never achieve the capacities of the demon. At any given stage of human knowledge, there will be many facts for which our best scientific knowledge will not provide any D-N explanation. In such situations we may see fit to resort to probabilistic inferences or probabilistic explanations or both. Under these circumstances the probabilities that are invoked reflect our ignorance. Explanations that do not qualify as D-N are simply incomplete; they do not represent bona fide explanations of a different type, such as inductive-statistical (I-S). When confronted by a putative explanation that is not D-N, the natural and legitimate response is to ask what is missing—not to seek a different model to characterize acceptable scientific explanations.

Although Carl G. Hempel gave the first precise, detailed formulation of the D-N model of scientific explanation, and defended it ably against many objections, he was not a partisan of explanatory deductivism. The classic paper by Hempel and Paul Oppenheim, "Studies in the Logic of Explanation" (1948), contains an explicit statement to the effect that not all legitimate scientific explanations fit the D-N model—instead, some are probabilistic or statistical (pp. 250–251). The task of analyzing such explanations and providing a precise model was, to be sure, not attempted in that paper; it was postponed until another time. The I-S model was first presented (though not under that name) in Hempel's "Deductive-Nomological vs. Statistical Explanation" (1962); an improved version was offered in his *Aspects of Scientific Explanation* (1965). In spite of these facts, I have often (even after 1962) encountered the belief that Hempel was committed to explanatory deductivism.

According to Hempel's theory of D-N explanation, the explanandum may be either a particular event or a general regularity. In either case the same model applies. The explanation is a valid deduction; at least one statement of a law occurs essentially in the explanans; and the conclusion states that the event (or fact) to be explained actually occurs (or obtains). Similarly, in the 1962 paper Hempel maintained that in the case of probabilistic or statistical explanation the explanandum may be either a particular event or a general statistical regularity—but there is a fundamental difference.

When the explanandum is a statistical generalization, the explanation may be deductive. For example, it might be wondered why a player who tosses a pair of standard dice twenty-four times has less than a fifty-fifty chance of getting double six at least once. It is reported that this problem bothered the Chevalier de Méré in the seventeenth century, and that Pascal was able to solve it by proving, with statistical generalizations about standard dice, that twenty-five throws are required to have better than a fifty-fifty chance for double six (Salmon 1967, p. 68). Although Hempel called attention to statistical explanations of this sort in the 1962 paper (p. 122), the formal model—known as deductive-statistical (D-S) explanation—was first presented three years later (Hempel 1965).

When, however, we explain some particular event or fact on the basis of statistical laws, the explanation cannot have the form of a deductive argument but must rather assume an inductive form. One of Hempel's most famous examples (1965, pp. 381–382 and 394–398) accounts for the quick recovery of John Jones from a streptococcus infection on the basis of the fact that penicillin was administered and the statistical regularity that almost all (but not all) such infections clear up promptly after treatment with penicillin. Explanations of this type have the inductive form often called "statistical syllogism"; they conform to the I-S model of explanation.

The explanatory deductivist can comfortably admit three kinds of explanation: explanation of individual events by subsumption under universal laws, explanation of universal regularities by subsumption under more general universal laws, and explanation of statistical regularities by subsumption under more general statistical laws. Indeed, to handle these three types, there is no need for any model beyond the D-N, for in all of them, some fact (particular or general) is explained by deduction from premises that include essentially at least one statement of a law. The D-S model as a separate entity is not needed.

The point at which the explanatory deductivist must take umbrage is when models of explanation—such as the I-S model—that characterize explanations as nondeductive arguments are introduced.[2] If there are bona fide scientific explanations that conform to the I-S model, that fact reveals a crucial limitation on deductivism.

Let us reconsider the streptococcus example. When Hempel first presented this case it was known that most streptococcus infections could be handled effectively by the administration of penicillin, but that some strains of these bacteria are penicillin resistant. If Jones is infected by a non-penicillin-resistant strain, then his prompt recovery after treatment is practically certain, though not absolutely certain. In response to this example, the deductivist can say that the "explanation" of the quick recovery on the basis of treatment with penicillin has explanatory value, though it is not a complete explanation. The "explanation" in terms of a non-penicillin-resistant strain of bacteria and treatment by penicillin is more complete, and hence possesses more explanatory value than the "explanation" in terms of treatment by penicillin alone. But it falls short of being a bona fide explanation. Most of us would agree that it is very probably incomplete, for there is good reason to suppose that further research will reveal additional factors that help to determine which victims of streptococcus infection by non-penicillin-resistant strains will recover quickly after treatment with penicillin and which will not.

Cases of this sort need not cause the deductivist any discomfort. The inferential deductivist may readily grant that many proffered arguments—ones that seem compelling to common sense—are actually enthymemes. Enthymemes are incomplete deductive arguments; it is possible to complete them by furnishing missing premises.[3] Similarly, the explanatory deductivist can, with impunity, accept explanations that conform to the I-S model, as long as they are regarded as incomplete D-N explanations. As such, these I-S explanations are literally enthymemes. In many such cases we do not yet have access to the additional true premises needed to transform the enthymeme into an acceptable D-N explanation, but we

the candy bar in terms of the insertion of the shilling if putting in the coin is sufficient for the result. This is the situation for machine *M*. Moreover, if putting in the coin is necessary but not sufficient, it cannot provide a D-N explanation of the emission of the candy bar. This characterizes machine *L*. To those of us who see a close relationship between causation and explanation, this outcome seems wrong. If one were to accept Mackie's account of causality and the deductivist's account of explanation, it would be necessary to conclude that putting the coin in machine *L causes* the candy bar to come out but does not *explain* its appearance, whereas putting the coin in machine *M explains* the appearance of the chocolate bar but does not *cause* it to emerge. This result is quite paradoxical.

The difficulty that arises in connection with machine *M*, it should be noted, strongly resembles a well-known problem for D-N explanation, namely, the problem of overdetermination. Consider a California ticky-tacky house built near the San Andreas Fault. If an earthquake measuring 7.0 or greater on the Richter scale is centered nearby the house will collapse. Likewise, if a tornado touches down right there the house will also collapse. One day a major earthquake does occur in that area and the house collapses. We have all the makings of a D-N explanation. However, the collapse of the house is not due to the earthquake, for a tornado knocks it down just before the earthquake occurs. In the case of machine *M*, it may be that the candy bar would have been delivered quite by chance, and thus that the insertion of the shilling had nothing to do with its appearance.

Since Mackie's machines—especially *L* and *M*—may seem rather artificial, let us consider a more realistic scientific example, one in which it is plausible to suppose that genuine quantum indeterminacy is involved. In 1972–73, Coffa and I signed up informally as lab partners in a course in which some of the landmark experiments in early twentieth-century physics were repeated (but not necessarily in the original way). As our first project we did a Compton scattering experiment, using a monochromatic gamma ray beam. It was necessary, of course, to detect scattered gamma ray photons. We constructed a detector consisting of a sodium iodide crystal, which scintillates when the scattered photons impinge upon it, and a photomultiplier tube, which detects the scintillations and sends a signal to a counter. Such detectors are not perfect.

Suppose, for the sake of argument, that our detector has an efficiency of .95 (a realistic value for commercially available detectors)—that is, it responds, on the average, to ninety-five out of a hundred photons impinging upon it. If it fails to detect some photons but never responds when no such photon is present, it seems that we would have little hesitation in explaining a detection event in terms of the incidence of a photon. If someone asked why the counter clicked, it would be appropriate to answer that a photon interacted with the sodium iodide crystal, which produced a flash of light that passed through the window of the photomultiplier tube and struck a metal surface, which in turn ejected a photoelectron. This electron initiated a cascade of secondary electrons, thus producing a brief pulse of electricity that activated the counter and made it click. Even if a small percentage of photons fail to activate the detector, we explain each response on the basis of an incident photon. It should be carefully noted that this explanation does not fit the D-N pattern, because the impinging of the photon on the device is a necessary but not sufficient condition for the occurrence of the click. This case corresponds to Mackie's machine L.

Suppose instead—again, for the sake of argument—that the detector sometimes produces a click when no scattered photon is present—for example, on account of a stray thermal electron in the photomultiplier tube—but never fails to detect an entering photon that was Compton-scattered into the detector. Again, it is realistic to suppose that five counts out of a hundred are spurious. It seems highly dubious that the deductivist, if asked to explain a given click, would be right in insisting that the story told in the preceding paragraph constitutes a bona fide explanation, even though it now fits the D-N pattern by providing a sufficient, but not necessary, condition for the click. In this case, if someone asked why the click occurred, it would seem far more appropriate to say that it might have been because a scattered photon entered the detector, producing a scintillation that then produced a cascade of electrons in the photomultiplier tube, or it might have been because of a thermal electron in the photomultiplier tube, though most probably it was on account of a scattered photon. This explanation obviously fails to fit the D-N pattern. It seems irrelevant to the explanation that every photon entering the tube produces a click. This case corresponds to Mackie's machine M.

Mackie's two indeterministic candy machines and the gamma

ray detectors resurrect—rather surprisingly, I think—the old problem about the explanatory value of necessary conditions. Consider Michael Scriven's syphilis/paresis example (1959). Paresis is one form of tertiary syphilis, and about one victim of latent syphilis out of four (untreated by penicillin) develops paresis. No one else ever contracts paresis. There is no known way to predict which victims of latent untreated syphilis will develop paresis and which will not. Scriven maintained that latent untreated syphilis explains paresis in those cases in which it occurs. Deductivists steadfastly claimed that we cannot explain paresis unless we can discover other factors that will serve to distinguish those victims of latent untreated syphilis who will develop paresis from those who will not. If the world is deterministic—at least with respect to syphilis and paresis—then there will be characteristics that make just this distinction, but if paresis occurs randomly and intrinsically unpredictably among victims of latent untreated syphilis, no such in-principle-discoverable characteristics exist. In this latter case, latent untreated syphilis is a necessary cause of paresis and there are no sufficient causes. Under these circumstances, I would suggest, untreated latent syphilis sometimes causes paresis and sometimes does not; but in all cases in which paresis occurs it is caused by latent untreated syphilis, and latent untreated syphilis explains the paresis. But for the latent untreated syphilis, the paresis would not have occurred.

The same sort of issue has arisen over the years in connection with the possibility of functional explanations. It might be asked, for example, why the jackrabbit—an animal that inhabits hot desert regions—has such large ears. The answer is that they act as devices for the dissipation of excess body heat. When the jackrabbit becomes overheated, it seeks the shade and dilates the many blood vessels in its ears. Heat is carried by the blood from the interior parts of the body to the ears and radiated into the environment. This provides a way for the animal to regulate the temperature of its body. I gather that elephants' ears function in a similar manner.

The trouble with functional explanations, from the standpoint of the deductivist, is that they furnish necessary conditions where the D-N model requires sufficient conditions. If we claim that regulation of body temperature explains the size of the jackrabbit's ears, then we can say that having big ears is a sufficient condition for regulation of body temperature, but we cannot say that regulation of body temperature is a sufficient condition for big ears, because there are other mechanisms of body-temperature control—for example, panting, perspiring, or nocturnal habits.

One response that is often given by deductivists consists of the claim that functional explanations of the foregoing sort are incomplete explanations. When we know more about the development of body-temperature control mechanisms we will find additional antecedent factors that determine that humans perspire, dogs pant, and mice avoid the heat of the day by staying in their shelters. It may turn out that, in the evolutionary context in which the large ears of the jackrabbit emerged, other mechanisms for the control of body temperature were not genetically available.

While I think that there is considerable hope in dealing with the syphilis/paresis example in terms of as-yet-unknown factors, I am far more skeptical about the feasibility of dealing with functional explanations in that way. One reason for this skepticism lies in the fact that some functional explanations involve characteristics that result from biological evolution. Since mutations play a vital role in such developments, we must not neglect the fact that these mutations arise by chance. This appeal to chance may sometimes be merely a euphemism for our ignorance. However, it is known that mutations can be produced by cosmic rays, and in such cases the fundamental quantum indeterminacy may well prevail.

There is another—far more basic—reason. Functional explanations are found in many scientific contexts, in social and behavioral sciences as well as evolutionary biology. Ethnologists, for example, explain social practices on the basis of their roles in fostering social cohesiveness. Freudian psychologists explain dreams in terms of wish fulfillments. In most, if not all, such cases, functional equivalents—alternative mechanisms that would achieve the same result—appear to be possible. I find no reason to suppose that such explanations are acceptable only on the supposition that it is possible in principle to show that one rather than another of these alternatives had to occur in the circumstances. Totemic worship of the wolf can be explained in terms of its social function even if totemic worship of the bear would have worked as well. Functional explanations do not cease to be explanations just because we cannot rule out functional equivalents.

We have looked at several examples in which there is a necessary cause but no sufficient cause, known or unknown, that invariably produces the event to be explained, and some in which there is a sufficient cause but no necessary cause, known or unknown, without which the event to be explained could not occur. Such examples necessarily involve some sort of indeterminism. In these indeterministic settings, it appears that necessary causes have at least some degree of explanatory force, but sufficient causes do not. This raises a serious question about the apparent plausibility of the deductivist's demand for sufficient causes in D-N explanations.

Consider the following example in contrast to the foregoing ones. On a stormy summer evening Smith's barn catches fire and burns to the ground. An insurance investigator wants to find out why. There are many possible causes of such a fire: lightning striking the barn, deliberate arson, a lighted cigarette discarded by a careless smoker, spontaneous combustion of green hay stored in the barn, and many others. The investigator establishes the fact that, even though there was a thunderstorm, lightning did not strike in the vicinity of the barn at the time the fire started; moreover, the barn was protected by adequate lightning rods. It is further established that no smoker was in the barn at the time. Careful examination of the remains establishes the absence of any incendiary substance such as gasoline. The burn pattern fits with the supposition that the cause was spontaneous combustion. This explanation satisfies the insurance investigator.

It would be natural at this point to ask what difference between the photon detector example and the burning barn example accounts for the fact (as I see it, anyhow) that a necessary condition has explanatory force in the former while a sufficient condition has explanatory force in the latter. One important difference is this. In the case of the barn it is plausible to assume that there always is a sufficient cause, and that different causes of burnt barns leave different traces in the effect. The photon detector does not have this characteristic. The click that results from a genuine photon-detection is utterly indistinguishable from the click that results from a spurious count.

One outstanding feature of such examples as the burnt barn is that they are assumed to be fully deterministic. Several possible sufficient conditions have to be considered, but (assuming no overdetermination) it is supposed that one and only one must have been present. Indeed, the presumption is that conditions surrounding this particular occurrence can be specified in enough detail to establish the existence of a unique necessary and sufficient cause. This situation is a deductivist's dream come true; the case fits the D-N model to a tee. When, however, we allow for the possibility of ineluctable indeterminacy the D-N model loses a good deal of its earlier appeal.

In this section we have discussed a very special aspect of indeterminism—one in which we did not need to make reference to any genuinely statistical laws. We found a significant limitation of explanatory deductivism, namely, its demand for sufficient causes in cases in which not they but rather necessary causes seem manifestly to have explanatory import. In the next section we shall deal with issues that arise directly from the statistical character of basic physical laws. Additional limits of deductivism will appear.

Scientific Explanation and Irreducibly Statistical Laws

If Laplacian determinism is fertile soil for explanatory deductivism, it might be supposed that the indeterministic context of twentieth-century physics would prove quite barren. On the standard Born-Pauli statistical interpretation of quantum mechanics, at least some of its basic laws are *ineluctably* statistical. Moreover, it has achieved important explanatory successes. Consider some of the most pressing problems that led to the discovery of quantum mechanics. Classical physics could not explain the energy distribution in the spectrum of blackbody radiation; quantum mechanics could. Classical physics could not explain the stability of the hydrogen atom or the discrete character of the hydrogen spectrum; quantum mechanics could. Similar remarks can be made about specific heats, the photoelectric effect, and radioactivity. Quantum mechanics thus seems to provide explanatory power of an irreducibly statistical kind. Modern physics therefore appears to mandate a theory of statistical explanation.

Even if it should turn out that quantum mechanics in the standard interpretation is incorrect and has to be replaced by a deeper, deterministic theory, it is surely advisable at this juncture to leave the door open to indeterminism and not to close off that possibility by some a priori fiat. If we have a model of statistical explanation, yet determinism is true after all, then at worst we have a model of scientific explanation that is not fundamental. If we have no model of statistical explanation, yet indeterminism is true, then we will have failed to comprehend a fundamental type of scientific explanation. It is obvious, in my opinion, that the second error would be much more serious than the first.

In an earlier section we discussed Hempel's example of the streptococcus infection, and we considered ways in which the deductivist might handle it. This example did not appear to present any insuperable difficulties. Not all of Hempel's classic examples can be handled as readily by the deductivist. Consider a particular ten-milligram sample of radon, a radioactive element with a half-life of 3.82 days (Hempel 1965, p. 392). The statistical law concerning the half-life of radon could be invoked to explain why, after 7.64 days, this sample contains about 2.5 milligrams of radon. Radioactive decay is one of the phenomena in which the ineluctably statistical character of quantum mechanics is displayed. To the best of our current knowledge, there is no strict law that determines just which nuclei will decay within a given period of time and which will remain intact. Indeed, the best current theory gives deep reasons for maintaining that there cannot be any deterministic explanation. Each atom has a fifty-fifty chance of decaying within any period of 3.82 days; that is the whole story. It is, consequently, impossible to deduce the statement that approximately 2.5 milligrams remain after 7.64 days from the laws and initial conditions given above.

As mentioned earlier, Hempel offered two models of statistical explanation, D-S and I-S. The radon example does not qualify as a D-S explanation, for it offers an explanation of the fact that this particular sample contains approximately 2.5 milligrams of radon at the end of a period of 7.64 days. Using the information given in the example, it would be possible to construct a D-S explanation of the fact that, in general, ten-milligram samples of radon almost always contain about 2.5 milligrams 7.64 days later. However, it is impossible to deduce from the information given anything about the radon content of this particular sample (beyond, perhaps, the conclusion that the value must be somewhere between zero and ten milligrams). Moreover, even if we agree with Hempel's analysis of this example as an instance of I-S explanation, it seems plausible to deny that it is epistemically relativized, and hence, an incomplete D-N explanation, for we have good reasons for believing that there are no additional factors, as yet unknown, that determine precisely which radon atoms will decay in 7.64 days, or precisely how many. Instead, it is a strong candidate to qualify as a bona fide complete I-S explanation.

Even philosophers who are uncommitted to determinism find basic difficulties with the I-S model. In his revised presentation of that model (1965) Hempel required that the explanandum be highly probable in relation to the explanans. One fundamental problem concerns the question how high is high enough. Take, for example, the explanation of the melting of an ice cube on the basis of the fact that it was placed and left in a tumbler of tepid water. Even in classical statistical mechanics we must admit that the melting is not a necessary consequence of the temperature of the surrounding water, for within the class of states of lukewarm water there is a minute fraction that will not eventuate in the melting of the ice cube within a given span of time. This is a consequence of the fact that the entropy in the system consisting of the water and the ice cube is overwhelmingly likely, but not absolutely certain, to increase. Because of the incredibly small difference between unity and the probability of melting, most philosophers will, I am confident, have little, if any, uneasiness in accepting this explanation as legitimate, even though strictly speaking it does not qualify as D-N. Hempel's example of radon decay would count as legitimate in the same fashion. These examples are in the same category as Richard Jeffrey's "beautiful cases," in which "the probability of the phenomenon to be explained is so high, given the stochastic law governing the process that produces it, as to make no odds in any gamble or deliberation" (1969, p. 27). As Jeffrey remarks, such cases surely provide "practical certainty"—though I am not sure what role that concept can play in the deductivist's scheme of things. A serious attempt to deal with this question has been made by J. W. N. Watkins (1984, pp. 242–246).

Let us consider the opposite extreme. It seems unlikely that anyone would admit a probability value in the closed interval from zero to one-half as a high probability; a minimal demand would be that the explanandum be more likely than not relative to the explanans. However, most people would find it hard to accept an I-S explanation of some event if the explanans renders it just a little more likely than not to occur. Few would be willing to agree that the sex of a newborn baby boy is explained in terms of the slight superiority of the statistical probability for boys over girls. Most people would have similar reluctance to admit an I-S explanation of the outcome heads on a toss of a coin, given that the coin has only a very slight bias toward heads—say, fifty-one to forty-nine.

The problem we confront involves a basic slippery slope. A probability virtually indistinguishable from one seems adequate for an I-S explanation; a probability scarcely above one-half seems clearly inadequate. If I-S explanations (in their full-blooded, unrelativized sense) are admissible at all, it is hard to see where to draw the line between probability values that are high enough and those that are not.

At various times I have invited consideration of examples in which there are two alternative outcomes having statistical probabilities of three-fourths and one-fourth respectively. One of these examples—taken from Hempel (1965, pp. 391–392)—involves a Mendelian genetic experiment on the color of blossoms of pea plants in a population having a certain genetic makeup, in which the probability of a red blossom is three-fourths and the probability of a white blossom is one-fourth. With admitted oversimplification, the occurrence of brown or blue eye colors in children of brown-eyed, heterozygous parents has the same probability distribution (Salmon 1985). Similarly, the probability of radioactive disintegration of an unstable nucleus—for example, one of Hempel's radon nuclei—within a period of two half-lives, or of its survival for at least that length of time, also exhibits the same probability distribution (Hempel 1965, p. 392; Salmon 1985).

It should be clearly noted that Hempel's use of these examples is quite different from mine. He does not discuss the explanation of a single instance of a red blossom or the decay of an individual nucleus. In the radon example, he specifies the size of the sample (ten milligrams); it contains about 3×10^{19} atoms. Obviously, as previously remarked, in a sample of this size the probability of getting approximately seventy-five percent decays in two half-lives is extremely close to one.[4]

Hempel's discussion of the distribution of red and white blossoms involves samples of unspecified size taken from a finite population of pea plants (1965, pp. 391–392). It therefore nicely illustrates the problem of the slippery slope. In a sample containing only one member, there is a probability of three-fourths that the distribution of red flowers will approximate the probability of red in the population, for a frequency of one is closer to three-fourths than is a frequency of zero. Is three-fourths large enough to qualify as a high probability? I shall return to that question shortly. In the meantime, it should be noted that, in a sample of ten, there is a probability of .42 that the frequency of red will be in the range .675 to .825—that is, within ten percent of .75. In a sample of fifty, the probability is .78 that the frequency will lie within that range; in a sample of one hundred, the probability is .92; in a sample of five hundred, the probability is virtually indistinguishable from one. The general situation is clear. Choose any degree of approximation you wish. By taking larger and larger samples, the probability that the frequency matches three-fourths within that degree of approximation may be made as close to unity as you wish. Therefore, if there is a probability, less than one, that qualifies as a high probability, we can find a finite sample size in which we will have a high probability of getting approximately three-fourths red blossoms. If your chosen degree of approximation is ten percent, and if any probability exceeding .9 is high enough for I-S explanation, then we can give an I-S explanation of the fact that the frequency of red in a sample of one hundred lies between .675 and .825, but we cannot give that kind of explanation of the fact that the frequency of red in a sample of fifty lies between thirty-four and forty-one. The question is how to draw the line between probabilities that are high enough for I-S explanation and those that are not in any nonarbitrary way.

It is interesting to note that almost no one seems to accept three-fourths as a sufficiently high value; nearly everyone seems to maintain that, if there is any explanation of the color of a single blossom (or the blossoms on a single plant), it must be in terms of the details of the chromosomal processes that causally determine the color. The reluctance to allow explanation of the statistical makeup of larger samples in terms of the probability distribution is not nearly as great.

There is another way to look at these cases in which the two outcomes have probabilities of three-fourths and one-fourth. It involves an important symmetry consideration. Given our knowledge of Mendelian genetics and the ancestors of the pea plants, it seems clear to me that we understand the occurrence of a white blossom just as well or just as poorly as we understand the occurrence of a red blossom. We can explain either red or white—whichever happens to occur—or we can explain neither.[5] Similarly, our knowledge of radioactive disintegration provides us with equally good understanding either of disintegration of a radon

nucleus within two half-lives or of survival intact for the same period.

The conclusion I have drawn from this symmetry argument (as well as many other considerations) is that the high-probability requirement is not a suitable constraint to impose on statistical explanations. My suggestion has been to adopt something akin to the statistical-relevance (S-R) model. Deductivists have been understandably reluctant to accept this suggestion, for not only does it reject the thesis that explanations are deductive arguments—it also rejects the thesis that explanations are arguments of any variety.

A frequent response to this symmetry argument is to deny that we can give explanations of any of these particular cases. What can be explained, many people say, is a statistical distribution in a class of individuals (see Watkins 1984, chap. 6, for a very clear account). As we have noted, Hempel introduced the genetic example to illustrate the explanation of the proportion of red blossoms in a limited sample of pea plants. He introduced the nuclear disintegration example to illustrate the explanation of the proportion of radon atoms in a particular sample that undergo transmutation. He fully realized that no deductive explanation can be given of the statistical distribution in any sample that is smaller than the population from which it is drawn. These examples were explicitly offered as instances of I-S explanation in which the probability of the explanandum relative to the explanans falls short of unity.

The symmetry consideration is rather far-reaching. For any event E that occurs with probability p, no matter how high—provided $p < 1$—there is a probability $(1 - p) > 0$ that E does not occur. For any unstable nucleus, the probability that it will disintegrate before n half-lives have transpired is $1 - 1/2^n$, which can be made arbitrarily close to one by choosing an n large enough. If this symmetry argument is cogent, it shows that there is no probability high enough for I-S explanations, because the larger p is, the smaller is $1 - p$. The symmetry argument says that we can explain the improbable outcome whenever it occurs if we can explain the highly probable outcome. This argument strikes me as sound.

A natural response of those who appreciate the force of the symmetry consideration is to deny the existence of genuine I-S explanations altogether. Instead of attempting to find refuge in sufficiently high probabilities they maintain that there are no statis-

tical explanations of individual events but only D-S explanations of statistical regularities. This means, of course, that there are no statistical explanations of frequency distributions in limited samples, no matter how large, provided they fall short of the total population. Since the D-S species of D-N explanations must appeal to statistical laws, and since such laws typically apply to potentially infinite classes, we can say, in principle, that there can be no D-S explanations of frequency distributions in finite samples.

The deductivist who takes this tack can readily agree that quantum mechanics does provide deductive explanations of all such phenomena as the energy distribution in the spectrum of blackbody radiation, the stability of the hydrogen atom, the discreteness of the hydrogen spectrum, the photoelectric effect, radioactivity, and any others that quantum mechanics has treated successfully. In each of these examples the explanandum is a general regularity, and it is explained by deductive subsumption under more general laws. Following this line the deductivist maintains that we can explain why $_{86}Rn^{222}$ has a half-life of 3.82 days, but we cannot explain the decay of a single atom, the survival of a single atom, or the decay of a certain proportion of radon atoms in a particular sample of that gas. We can explain why, in the overwhelming majority of ten-milligram samples of radon, approximately 2.5 milligrams remain at the end of 7.64 days. We can remark *that* the particular sample described by Hempel behaved in the typical way, but we cannot explain *why* it did so.

In the preceding section, we discussed two types of hypothetical gamma ray detectors, each of which was subject to a particular kind of inaccuracy. In the real world we should expect inaccuracies of both kinds to occur within any such detecting devices. A gamma ray detector generally fails to respond to every photon that impinges upon it, and it may also give spurious counts. Thus, realistically, the impinging of a photon is neither a necessary nor a sufficient condition for the click of the counter. In addition, the probabilities in this example are modest—they are *not* for all practical purposes indistinguishable from one or zero. The discrepancies from the extreme values do make a significant difference. Under these circumstances we must conclude, I believe, that it is impossible in principle to provide a deductive explanation of such a simple phenomenon as the click of a counter. If any explanation of such an individual event is possible it will have to fit a nondeductive pattern

such as Hempel's I-S model, my S-R model, or Railton's deductive-nomological-probabilistic (D-N-P) model.[6]

The Profit and the Price

If indeterminism is true there will be types of circumstances C and types of events E such that E sometimes occurs in the presence of C and sometimes does not. We can know the probabilities with which E happens or fails to happen, given C, but we cannot, even in principle, know of any circumstances in addition to C that fully determine whether E happens or not. Even if the probability of E given C is high, many philosophers will reject the notion that an explanation of a particular case of E can be given. It may, of course, be possible to give a D-N explanation of the statistical regularity that E follows C with a certain probability p.

Georg Henrik von Wright (1971, p. 13) has argued that, in such cases, we cannot have an explanation of E, given that E occurs, for we can still always ask why E occurred in this case but fails to occur in others; Watkins (1984, p. 246) expresses a similar view. Circumstances C, von Wright holds, may explain why it is reasonable to expect E, but they cannot explain why E occurred. Wolfgang Stegmüller (1973, p. 284) responds to the same situation by observing that, if we claim that E is explained by C when E occurs, then we must admit that C sometimes explains E and sometimes does not. Since, by hypothesis, given C, it is a matter of chance whether E occurs or does not, it becomes a matter of chance whether C explains anything or not. Stegmüller finds this conclusion highly counterintuitive. While he admits the value of what he calls statistical deep analysis—something that is closely akin to the I-S and S-R models—he denies that it qualifies as any type of scientific explanation.

The counterintuitive character of statistical explanation becomes more dramatic if we invoke the symmetry consideration discussed in the preceding section. Given that E follows C in the vast majority (say ninety-five percent) of cases, E fails to happen a small minority (five percent) of the time. If, for example, we have an electron with a certain amount of energy approaching a potential barrier of a certain height, there will be a probability of .95 that it will tunnel through and a probability of .05 that it will be reflected back. If we claim to understand why the electron got past the barrier in one

case, then we must understand just as well why in another case with the same conditions it was turned back. The explanatory theory and the initial conditions are the same in both cases. Thus, it must be admitted, circumstances C sometimes explain why the electron is on one side of the barrier and sometimes why it is on the other side. Circumstances C are called upon to explain whatever happens.

There is a fundamental principle that is often invoked in discussions of scientific explanation. Watkins (1984, pp. 227–228) explicitly adopts it; Stegmüller calls it "Leibniz's principle" (1973, pp. 311–317); D. H. Mellor refers to it as "adequacy condition S" (1976, p. 237). I shall call it "Principle 1" (Salmon 1984, p. 113). It might be formulated as follows:

> It is impossible that, on one occasion, circumstances of type C adequately explain an outcome of type E and, on another occasion, adequately explain an outcome of type E' that is incompatible with E.

It is clear that D-N explanations never violate this condition; from a given consistent set of premises it is impossible to derive contradictory conclusions. It is clear, in addition, that I-S explanations complying with the high-probability requirement satisfy it, for on any consistent set of conditions, the sum of the probabilities of two inconsistent outcomes is, at most, one. Since any high probability must be greater than one-half, at most one of the outcomes can qualify. As we noted in the preceding section, however, it is difficult to see how the high-probability requirement can be maintained without an extreme degree of arbitrariness. If that result is correct, then it appears that Principle 1 draws the line constituting the limit of explanatory deductivism. In making this statement I am claiming that anyone who rejects explanatory deductivism must be prepared to violate Principle 1. The question then becomes, Is it worth the price?

In an earlier work (Salmon 1984, pp. 112–120), I argued at some length and in some detail that violation of Principle 1 does not have the dire consequences that are generally assumed to ensue. If we pay sufficient attention to avoidance of ad hoc and vacuous explanatory "laws," we can disqualify the undesirable candidates for scientific explanations without invoking Principle 1. Abandonment of that principle need not be an open invitation to saddle science with pseudo-explanations.

In the same place I argued, on the basis of several striking examples, that twentieth-century science contains statistical explanations of the nondeductive sort. One of these was a Mendelian genetic experiment on eye color in fruit flies conducted by Yuichiro Hiraizumi in 1956. Although it looked in advance just about the same as the genetic experiments mentioned above, it turned out that in a small percentage of matings, the statistical distribution of eye colors was wildly non-Mendelian, while in the vast majority of matings the distribution was just what would be expected on Mendelian principles. A new theory was needed to provide an explanation of the exceptional outcomes in the particular matings that were observed. A possible explanation would attribute the exceptional distributions to chance fluctuations under standard Mendelian rules; that explanation could not, however, be seriously maintained. The preferred explanation postulates "cheating genes" that violate Mendel's rules (see Cohen 1975).

Another example involved the spatial distribution of electrons bouncing off of a nickel crystal in the Davisson-Germer experiment. The periodic character of the distribution, revealing the wave aspect of electrons, was totally unanticipated in the original experiment and demanded theoretical explanation. That observed pattern involved a finite number of electrons constituting a limited sample from the class of all electrons diffracted by some sort of crystal.

The deductivist can reply that in both of these cases what is sought is a statistical theory that will explain, in general, the occurrence of statistical distributions in limited samples of the types observed. When the mechanism of "cheating genes" is understood we can explain why occasional matings will produce results of the sort first observed by Hiraizumi. Similarly, quantum mechanics explains why, in general, electron diffraction experiments will very probably yield periodic distributions. In both cases, the deductivist might say, theoretical science shows us how such occurrences are possible on some ground other than the supposition that they are incredibly improbable chance fluctuations. The deductivist can maintain, in short, that all such statistical explanations in pure science are of the D-S type.

Consider another example. Suppose that an archaeologist, studying a particular site, comes across a piece of charcoal that has a concentration of C^{14} in its carbon content that is about half the concentration found in trees growing there at present. Since the

half-life of C^{14} is 5,568 years, the archaeologist explains the difference in C^{14} concentration by supposing that the tree from which this charcoal came was felled, and consequently ceased absorbing CO_2 from the atmosphere, about 5,568 years ago. Although there are many potential sources of error in radiocarbon dating, it is not seriously supposed that the tree was felled 2,784 years ago, and that by chance its C^{14} decayed at twice the normal rate. This example—like the melting ice cube, the radon decay, and the Davisson-Germer experiment—qualifies as a Jeffrey-type "beautiful" case.

The deductivist might, it seems to me, reply to the "beautiful" cases that, strictly speaking, we can furnish only D-S explanations of the statistical distributions of melting of ice cubes, the behavior of diffracted electrons, and rates of decay of radioactive isotopes. However, in the "beautiful" cases the statistical distributions show that a different outcome is so improbable that, though it is not physically impossible, we can be confident that neither we nor our ancestors nor our foreseeable descendents have ever seen or will ever see anything like it—anything, that is, as egregiously exceptional as an ice cube that does not melt in tepid water or a large collection of radioactive atoms whose rate of disintegration differs markedly from the theoretical distribution. We may therefore be practically justified in treating the theoretical statistical relationship as if it were a universal law.[7]

Archaeology, unlike many areas of physics perhaps, is usually concerned with particulars—particular sites, particular populations, particular artifacts, and particular pieces of charcoal. In the preceding example the archaeologist could invoke a precise physical law concerning radioactive decay to explain the C^{14} content of a particular piece of charcoal. In other cases no such exact general laws are available; at best there may be vague statistical relationships to which an appeal can be made. For instance, archaeologists working in the southwestern United States would like to find out why one particular habitation site (Grasshopper Pueblo) was abandoned at the end of the fourteenth century, and more generally, why the same thing happened all over the Colorado Plateau within a relatively short period of time (see Martin and Plog 1973, pp. 318–333). Various factors such as overpopulation followed by drought can be adduced by way of explanation, but there is no real prospect of a D-N explanation in either the more restricted or the more general case. At best, any explanation will be probabilistic.

To examples of this sort the deductivist can readily respond that they involve obviously incomplete explanations. This point can hardly be denied. Certainly there are as yet undiscovered factors that contribute to the explanation of the phenomenon in question. The same kind of response is often made to examples of another type—namely, controlled experiments. Such cases occur very frequently, especially in the biological and behavioral sciences. Consider one well-known instance.

In the late 1970s some Canadian researchers studied the relationship between bladder cancer and saccharin in laboratory rats (see Giere 1984, pp. 274–276). The experiment involved two stages. First, large quantities of saccharin were added to the diet of one group ($n = 78$), while a similar group ($n = 74$) were fed the same diet except for the saccharin. Seven members of the experimental group developed bladder cancer; one member of the control group did so. The null hypothesis—that there is no genuine association between saccharin and bladder cancer—could be rejected at the .075 level. This result was not considered statistically significant.

The second stage of the experiment involved the offspring of the original experimental and control groups. In each case the offspring were fed the same diets as their parents; consequently, the members of the second-generation experimental group were exposed to saccharin from the time of conception. Fourteen members of this group ($n = 94$) developed bladder cancer, while none of the members of the second-generation control group did. This result is significant at the .003 level. It was taken to show that there is a genuine positive association between ingestion of saccharin and bladder cancer among rats. The difference between the incidence of bladder cancer in the experimental and control groups is explained not as a chance occurrence but by this positive correlation, which is presumed to be indicative of some sort of causal relationship.

What can we say about this explanation? First, if no such positive correlation exists, the proffered explanation is not a genuine explanation. Second, if the positive correlation does exist, it is clear that, by chance, on some occasions, no statistically significant difference will appear when we conduct an experiment of the type described. The deductivist must, consequently, reject the foregoing explanation, because the explanandum—the difference in frequency between the two groups—cannot be deduced from the explanans. The deductivist can, in this case, reiterate the response to the pueblo

abandonment example; namely, the statistical character of the explanation arises from its incompleteness. If we possessed complete information it would be possible to deduce, for each rat, whether it contracts bladder cancer or not.

Nevertheless, it seems to me, the claim that all such explanations are necessarily incomplete insofar as they fall short of the D-N model is extreme. If it arises from the assumption that all such cases are absolutely deterministic—and that all appeals to probability or statistics simply reflect our ignorance—then it is based upon a gratuitous metaphysics that appears to be incompatible with contemporary physical science. If we do not have an a priori commitment to determinism, there is no reason to deny that indeterminacy arises in the domains of the biological or behavioral sciences. At the same time, if indeterminacy does occur, why should we withhold the appellation "complete explanation" from an explanation that cites all factors that are statistically or causally relevant to the occurrence of the explanandum event?[8]

In an earlier work (Salmon 1984), I maintained that pure theoretical science includes nondeductive statistical explanations. In the present context, *for the sake of argument*, I am prepared to relinquish that claim. Let us therefore agree—for now—that all of the statistical explanations that occur in theoretical science are either D-S explanations of statistical regularities or incomplete D-N explanations of particular facts. For purposes of the present discussion I want to give deductivism the benefit of the doubt. I will therefore grant that the deductivist can—by treating every statistical explanation as either D-S or incomplete D-N—avoid admitting nondeductive statistical explanations as long as the discussion is confined to the realm of pure science. With respect to these examples, I know of no knockdown argument with which to refute the deductivist claim. But I shall try to show that the price the deductivist must pay is still exorbitant. It requires relinquishing the capacity to account for explanations in the more practical context of applied science.

Explanation in Applied Science

I am not prepared to concede, even for the sake of argument, that applied science can dispense with nondeductive statistical explanations. Granted that many explanations encountered in practi-

cal situations may reasonably be regarded as I-S explanations that are incomplete D-N explanations, there are others that defy such classification. Let us look at a couple of examples.

When Legionnaires' disease was first diagnosed in 1976, it was found that every victim had attended an American Legion convention in Philadelphia, and that all of them had stayed at one particular hotel. In the population of individuals attending that convention, residence at that hotel was a necessary but by no means sufficient condition of contracting the disease. Later, after the bacillus responsible for the disease had been isolated and identified, it was found that cooling towers for air conditioning systems in large buildings sometimes provide both a favorable environment for their growth and a mechanism to distribute them inside of the building. In this case, as well as in subsequent outbreaks in other places, only a small percentage of the occupants of the building contracted the disease. Since quantum fluctuations may lead to large uncertainties in the future trajectories of molecules in the air, and to those of small particles suspended in the atmosphere, I believe it quite possible that there is, even in principle, no strictly deterministic explanation of which bacteria entered which rooms and no strictly deterministic explanation of which people occupying rooms infested with the bacteria contracted the disease. Nevertheless, for purposes of assigning responsibility and taking preventive steps in the future, we have an adequate explanation of the disease in this very limited sample of the population of Americans in the summer of 1976. It is a nondeductive statistical explanation that, admittedly, may be incomplete. There is, however, no good reason to suppose that it can, even in principle, be transformed by the addition of further relevant information into a D-N explanation of the phenomenon with which we are concerned (see Salmon 1984, p. 212).

Eight soldiers, out of a group of 2,235 who participated in operation Smoky in 1957, witnessing the detonation of an atomic bomb at close range, subsequently developed leukemia. The incidence—which is much greater in this group than it is in the population at large—is explained by the high levels of radiation to which they were exposed (Salmon 1978, pp. 688–689, and n. 15). Because leukemia occurs with nonzero frequency in the population at large, it is possible, but not likely, that the high incidence of leukemia in this sample of the population was due to a chance fluctuation rather than an increased probability of leukemia as a result of exposure

to a high level of radiation. From a practical standpoint, the fact to be explained is the high incidence in this particular sample, not the incidence among all people who ever have been or will be exposed to that amount of radiation; thus, it would be a mistake to construe the explanation as D-S. It is the occurrence of leukemia in this particular sample that has obvious importance in deciding such questions as whether these soldiers should receive extra compensation from the federal government. I am by no means certain that there is, in principle, no deterministic explanation of the onset of leukemia; however, because of the crucial involvement of radiation, it is not implausible to suppose that certain aspects are irreducibly statistical. In that case it would be impossible in principle to provide a D-N explanation of this phenomenon.

When it comes to the question of explaining the individual cases of leukemia, we must admit that we know of no factors that are either necessary or sufficient. Any given member of the group might have contracted leukemia even if he had not participated in operation Smoky, and the vast majority of those who were involved did not contract leukemia. It is quite possible that other relevant factors bearing on the occurrence of leukemia were operative, but there is no guarantee that they add up to either sufficient or necessary conditions.

When we try to explain some occurrence, we may have any of several purposes. First, we may be seeking purely intellectual understanding of the phenomenon in question. Depending on one's philosophical biases, such understanding may result from finding the causes that produced the phenomenon or from subsuming it under a universal law. When there are no strict deterministic causes by which to account for it, or when there are no universal laws, we may be willing to settle for knowledge of the frequency with which *events of that type* are produced under specific conditions. The deductivist can accept this kind of understanding as the product of D-N explanation (recalling that D-S explanation is one of its subtypes). Such explanations can be attributed to pure science.

Second, when the occurrence to be explained is undesirable, we may wish to understand why it happened in order to take steps to prevent such occurrences in the future. Our practical purposes will be served if there is a necessary cause that is within our power to eliminate—for example, paresis would be prevented if syphilis were eradicated or treated with penicillin. Failing a strict necessary con-

dition, our practical purposes will be served if we can find conditions—again within our power to control—whose elimination will reduce the frequency with which the undesirable occurrence takes place. Finding that automobile accidents on snow-covered roads occur probabilistically as a result of inadequate traction, we see that accidents of this type can be reduced—though not completely eliminated—through the use of adequate snow tires.

Third, if the occurrence in question is one we consider desirable, we may seek to understand why it happened in terms of sufficient causes. If sufficient causes that are under our control exist, they may be brought about in order to produce the desired result. For example, we can explain why a satellite remains at a fixed location above the earth in terms of the radius of the geosynchronous orbit (on the basis of Kepler's third law). A satellite can be placed in a geosynchronous orbit by boosting it, via rockets, to the specified altitude above the earth (about 22,300 miles) and injecting it into orbit.

Fourth, we often try to understand a desirable result in terms of circumstances that are necessary to its occurrence. In such cases we may discover a necessary condition that is absent. For example, a certain amount of water is required if various crops—such as corn, hay, and cotton—are to flourish. In desert areas irrigation is practiced. Adequate water is not sufficient to ensure good crops; if the soil lacks certain nutriments, such as nitrogen, the crops will not be healthy. But without the required water, other steps, such as fertilization or crop rotation, will not yield bountiful harvests.

Fifth, explanations are sometimes sought in order to assign moral or legal responsibility for some happening—very often a harmful result. Situations of this sort may well provide the strongest case against deductivism in the realm of applied science. Operation Smoky is a good example. To ascertain whether the U.S. Army is responsible for the eight cases of leukemia among the soldiers who participated in that exercise, we want to determine whether exposure to intense radiation explains these cases of leukemia. In order to answer that question we need a general statistical law connecting leukemia with exposure to radiation. This law is a required component of the explanans. We are *not* trying to explain some general statistical regularity; we are trying to explain *these particular cases* of leukemia. We know of no universal laws that would make it possible to explain these particular instances deductively, and we have no reason to suppose that any such universal regularity exists unbeknownst to us.

At this juncture the deductivist would seem to have three possible rejoinders. First, he or she might simply deny that we have explanations of such phenomena as these particular cases of Legionnaires' disease and leukemia. This tack, it seems to me, runs counter to well-established and reasonable usage. It is commonplace, and unobjectionable, to maintain that we can explain the occurrence of diseases even when we have no prospects of finding sufficient conditions for a victim to contract it.

Second, the deductivist might insist that phenomena of this sort always do have sufficient causes, and that they are amenable to D-N explanation. The explanations we can actually give are therefore to be viewed as partial D-N explanations; they are not completely without practical value as they stand, but they are not genuine explanations until they have been completed. Whoever subscribes to this stubborn metaphysical dogmatism deserves the title "deductive chauvinist pig."

Third, the deductivist might claim that our nondeductive "explanations" are partial explanations even though, in some cases, it may be impossible in principle to complete them on account of the nonexistence of suitable universal laws. Such partial explanations, it might be maintained, have practical value even though they fall short of the ideals of explanation in the context of pure science.

There is a temptation to try to convict this third response as incoherent on the ground that "partial explanation" makes no sense where there is no possibility in principle of having complete explanations. Yet, it seems to me, that rejoinder would be philosophically unsound. If the relative sizes and distances of the sun, moon, and earth were just a little different there might be no such thing as a total eclipse of the sun; nevertheless, there would be nothing strange in talking about partial eclipses and in assigning degrees of totality.

The appropriate strategy might rather be to accept this third move, pointing out that it hardly qualifies as deductivism. If the concept of partial explanation is to be serviceable, there must be standards in terms of which to judge which partial explanations are strong and which weak, which are useful and which useless. Requirements akin to Hempel's maximal specificity (1965, pp. 399–400) or my maximal homogeneity (1984, p. 37) would be needed to block certain kinds of partial explanations. In short, the deductivist would need to develop a theory of partial explanation that would be a direct counterpart of Hempel's I-S model, my S-R model, my statistical-causal concept, or any of several others. If the deductivist accepts the fact that something that is not a deductive argument and cannot possibly be made into a deductive argument can nevertheless be an acceptable partial explanation, it seems to me that he or she has given up the deductivist viewpoint, and is simply substituting the expression "partial explanation" for the term "statistical explanation" as it is used by many philosophers who have rejected the deductivist viewpoint.

Conclusions

In this paper we visited explanatory deductivism in the context of Laplacian determinism, and we found it very much at home there. However, since we now have strong reasons to believe that our world does not conform to the deterministic model, we found it necessary to revisit explanatory deductivism in the modern context, where, quite possibly, some of the basic laws of nature are irreducibly statistical. Although explanatory deductivism does not reside as comfortably here, evicting it, we found, is no easy matter—as long as we confine our attention to pure science. The claim that every statistical explanation is either a D-S explanation of a statistical regularity or an incomplete D-N explanation of a particular fact proves difficult to dislodge.

When we turn our attention to applied science, however, the situation is radically different. Explanatory deductivism does not do justice to explanations in practical situations. An interesting parallel emerges. As I argue in "Rational Prediction" (chap. 3 in this volume), the most decisive argument against inferential deductivism arises in connection with the use of scientific knowledge to make predictions that serve as a basis for practical decisions. Both types of deductivism are unsuited for the practical realm.

Even in the realm of pure science, it seems to me, both forms of deductivism are untenable. Inferential deductivism fails to allow for predictions—such as the claim that our expanding universe will eventually begin an era of contraction that will lead to a "big crunch"—which have no practical import but a great deal of intellectual fascination. Explanatory deductivism encounters several difficulties. One that has emerged in this paper concerns the relations between sufficient and necessary conditions. In the second section we looked at cases—Mackie's candy machines and Coffa-Salmon photon detectors—in which a conflict arises between the D-N demand for sufficient conditions to explain what happens and the intuitive demand for causal explanations, where the cause in question is a necessary but not sufficient condition.[9] Thus, there arises a serious tension between the deductivistic conception of scientific explanation and the causal conception even in the realm of pure science.

In my earlier writing (Salmon 1984, esp. chaps. 1, 4, and 9), I attempted to compare and contrast several fundamental conceptions of scientific explanation, including the deductivistic and causal conceptions, in considerable detail.[10] In the context of Laplacian determinism they are virtually equivalent, and there is not much reason to prefer one to the other. In the modern context, in which at least the possibility of indeterminism must be taken seriously, the two conceptions diverge sharply. According to the causal conception, we explain facts (general or particular) by exhibiting the physical processes and interactions that bring them about. Such mechanisms need not be deterministic to have explanatory force; they may be irreducibly statistical. Causality, I argue, need not be deterministic; it may be intrinsically probabilistic. The benefit we obtain in this way is the recognition that we can provide scientific explanations of particular events that are not rigidly determined by general laws and antecedent conditions. As I argued in the fifth section above, "Explanation in Applied Science," the availability of such explanations is required for the application of science in practical situations; it also seems to be faithful to the spirit of contemporary pure science. The notion that we can explain only those occurrences that are rigidly determined is a large and unneeded piece of metaphysical baggage.

The price we pay for the claim that phenomena that are not completely determined can be explained is the abrogation of Principle 1. As I have argued at some length (Salmon 1984, pp. 113–120); Salmon 1985), the price is not too high. Principle 1 is, I believe, the explanatory deductive chauvinist's main bludgeon. Once it has been rendered innocuous the chief appeal of explanatory deductivism is removed.

Notes

At the 1980 Workshop I presented a paper, entitled "Causal and Statistical Relevance," as a preview of many discussions carried out more extensively and systematically in Salmon 1984. That paper was rendered otiose upon publication of the book. The present paper was written in 1985, expressly for this volume, in special recognition of Alberto Coffa's important contributions to the discussion of statistical explanation.

I should like to thank Philip Kitcher—a self-declared deductive chauvinist—for stimulating and helpful comments on an earlier draft of this paper.

1. The most cogent argument against Popper's inferential deductivism is given in my "Rational Prediction" (reprinted as chap. 3 of this volume). An exchange with John Watkins on this topic occurs in Salmon 1968*a*, pp. 25–29; Watkins 1968; and Salmon 1968*b*, pp. 95–97. A somewhat different slant is given in Salmon 1967, pp. 21–27 and 114–115.

2. Even worse for the deductivist is the statistical-relevance (S-R) pattern of scientific explanation. Explanations conforming to that model not only fail to be deductive; in addition they fail to qualify as arguments of any sort.

3. Bertrand Russell is the most distinguished proponent of this form of inferential deductive chauvinism; see Salmon 1974.

4. According to Watkins (1984, p. 243), the probability that the proportion of undecayed atoms lies within four percent of .25 is greater than $1 - 10^{-(10^{15})}$ when $n > 10^{19}$.

5. Critics tend to agree, but they insist that such understanding is possible only through knowledge of the chromosomal processes; see, for example, Kitcher 1985, p. 634, and van Fraassen 1985, pp. 641–642.

6. Railton's D-N-P model (1978) has the great virtue of demanding reference to the mechanisms that bring about such indeterministic results.

7. See Watkins 1984, pp. 242–246, for a detailed analysis of the "beautiful cases."

8. Philip Kitcher (1985, p. 633) suggests that statistical explanations of particular events be considered incomplete, not on the ground that nature can or must furnish additional explanatory facts but on the ground that the explanatory value of such explanations falls short of that of D-N explanations. According to his terminology, explanations that are irreducibly statistical are incomplete, not because of our epistemic shortcomings but because nature does not contain additional factors in terms of which to render them "ideally complete." I find it more natural to speak of complete nondeductive or complete statistical explanations.

9. In Salmon et al. 1971, pp. 58–62, I distinguish between relevant and

irrelevant necessary conditions and between relevant and irrelevant sufficient conditions. Irrelevant conditions of both kinds lack explanatory import. In addition, I discuss the statistical analogues of necessary and sufficient conditions; necessary and sufficient conditions are simply limiting cases of these statistical relationships. It is argued that both the statistical analogues of sufficient conditions and those of necessary conditions have explanatory import, but only if they are statistically relevant to the explanandum. Because these statistical relationships play no role in the deductivist account, I have not discussed them in the text of this paper. Nevertheless, it seems to me, the fundamental answer to the question whether sufficient or necessary conditions have explanatory import should be based on the relevancy relations offered in that discussion.

10. In Salmon 1984, I distinguish three major conceptions of scientific explanation—namely, epistemic, modal, and ontic. The view that all scientific explanations are arguments, either deductive or inductive, is identified as the inferential version of the epistemic conception; the doctrine that all explanations are deductive arguments represents the modal conception. The causal conception of scientific explanation is a version of the ontic conception.

References

Coffa, J. Alberto. 1974. "Hempel's Ambiguity," *Synthese* 28:141–164.

Cohen, S. N. 1975. "The Manipulation of Genes," *Scientific American* 233:24–33.

Giere, Ronald N. 1984. *Understanding Scientific Reasoning*, 2d ed. New York: Holt, Rinehart and Winston.

Hempel, Carl G. 1962. "Deductive-Nomological vs. Statistical Explanation." In H. Feigl and G. Maxwell, eds., *Minnesota Studies in the Philosophy of Science*, vol. 3, pp. 98–169. Minneapolis: University of Minnesota Press.

———. 1965. *Aspects of Scientific Explanation and Other Essays in the Philosophy of Science*. New York: The Free Press.

Hempel, Carl G., and Paul Oppenheim. 1948. "Studies in the Logic of Explanation," *Philosophy of Science* 15:135–175. Reprinted, with a "Postscript," in Hempel 1965, pp. 245–295.

Jeffrey, Richard C. 1969. "Statistical Explanation vs. Statistical Inference." In N. Rescher, ed., *Essays in Honor of Carl G. Hempel*, pp. 104–113. Dordrecht: Reidel. Reprinted in Salmon et al. 1971, pp. 19–28.

Kitcher, Philip. 1985. "Two Approaches to Explanation," *Journal of Philosophy* 82:632–639.

Mackie, J. L. 1974. *The Cement of the Universe*. Oxford: Clarendon Press.

Martin, Paul S., and Fred Plog. 1973. *The Archaeology of Arizona.* Garden City, N.Y.: Doubleday/Natural History Press.

Mellor, D. H. 1976. "Probable Explanation," *Australasian Journal of Philosophy* 54:231–241.

Railton, Peter. 1978. "A Deductive-Nomological Model of Probabilistic Explanation," *Philosophy of Science* 45:206–226.

———. 1980. *Explaining Explanation: A Realist Account of Scientific Explanation and Understanding.* Ph.D. dissertation, Princeton University. Ann Arbor, Mich.: University Microfilms International.

Salmon, Wesley C. 1967. *The Foundations of Scientific Inference.* Pittsburgh, Pa.: University of Pittsburgh Press.

———. 1968a. "The Justification of Inductive Rules of Inference." In I. Lakatos, ed., *The Problem of Inductive Logic,* pp. 24–43. Amsterdam: North-Holland.

———. 1968b. "Reply." In I. Lakatos, ed., *The Problem of Inductive Logic,* pp. 74–97. Amsterdam: North-Holland.

———. 1974. "Russell on Scientific Inference; or, Will the Real Deductivist Please Stand Up?" In G. Nakhnikian, ed., *Bertrand Russell's Philosophy,* pp. 183–208. London: Gerald Duckworth.

———. 1978. "Why Ask, 'Why?'? An Inquiry Concerning Scientific Explanation," *Proceedings and Addresses of the American Philosophical Association* 51:683–705.

———. 1981. "Rational Prediction," *British Journal for the Philosophy of Science* 32:115–125. Reprinted as chap. 3 of this volume.

———. 1984. *Scientific Explanation and the Causal Structure of the World.* Princeton, N.J.: Princeton University Press.

———. 1985. "Scientific Explanation: Three Basic Conceptions." In P. Asquith and P. Kitcher, eds., *PSA 1984,* vol. 2, pp. 293–305. East Lansing, Mich.: Philosophy of Science Association.

Salmon, Wesley C. et al. 1971. *Statistical Explanation and Statistical Relevance.* Pittsburgh, Pa.: University of Pittsburgh Press.

Scriven, Michael. 1959. "Explanation and Prediction in Evolutionary Theory," *Science* 130:477–482.

Stegmüller, Wolfgang. 1973. *Probleme und Resultate der Wissenschaftstheorie und Analytischen Philosophie,* Band 4, Studienausgabe Teil E. Berlin and New York: Springer-Verlag.

van Fraassen, Bas C. 1985. "Salmon on Explanation," *Journal of Philosophy* 82:639–651.

von Wright, Georg Henrik. 1971. *Explanation and Understanding.* Ithaca, N.Y.: Cornell University Press.

Watkins, J. W. N. 1968. "Non-inductive Corroboration." In I. Lakatos, ed., *The Problem of Inductive Logic,* pp. 61–66. Amsterdam: North-Holland.

———. 1984. *Science and Scepticism.* Princeton, N.J.: Princeton University Press.

6. A Nondeductivist Approach to Theoretical Explanation

Frederick Suppe

> It is the part of wisdom to ask not why, but how
> events happen.
>
> —André Gide

How do the limits of deductivism affect our philosophical under-standing of scientific explanation? Most philosophical discussions of explanation over the last half century have been dominated by Carl G. Hempel's contention that explanations are to be construed as deductive arguments. One of my main contentions in this paper is that the recent history of philosophical disputes over explanation strongly suggests that philosophical understanding of scientific ex-planation lies outside the limits of deductivism—that the most promising approach to understanding explanation will be non-deductive. Arguing this is one task of the first section of the paper, "Philosophical Analyses of Explanation"; its other task is to ferret out of the literature on explanation positive suggestions for devel-oping our replacement, nondeductivist analysis. The remaining sec-tions of the paper are concerned with presenting a nondeductivist analysis of theoretical explanation.

Although most philosophical analyses of explanation focus on the use of *laws* to explain specific events, one reason for confining my attention here to *theoretical explanations* is that in actual scien-tific practice they are far more important and central. Wesley Sal-mon makes the point as follows:

> Arguments by Greeno and others have convinced me that explanations of particular events seldom, if ever, have genuine scientific import (as opposed to practical value), and that explanations which are scientifi-cally interesting are almost always explanations of classes of events. This leads to the suggestion . . . that the goodness or utility of a scientific explanation should be assessed with respect to its ability to account for entire classes of phenomena, rather than by its ability to deal with any particular event in isolation.[1]

In fact there are two distinct sorts of theoretical explanations that explain entire classes of phenomena. One sort involves the use of a theory to explain another law or theory—as, for example, when Newtonian theory is used to explain Kepler's laws; this sort has been discussed by Hempel and others and is closely associated with intertheoretic reduction. The other sort, with which I am solely concerned in this paper, involves the use of theories to explain events; we will see that such theoretical explanations invariably account for entire classes of phenomena.

A second reason for focusing our attention on theoretical explanation here stems from our desire to present a nondeductivist analysis of scientific explanation. By the late 1960s a number of authors, including Evert Beth, Patrick Suppes, Bas van Fraassen, and myself, had concluded that the deductivist axiomatic approach used by positivists to analyze scientific theories was untenable, and we had attempted to develop nondeductivist analyses of the structure of scientific theories. The analyses we produced collectively have come to be known as the semantic conception of theories. The strategy of this paper is to exploit the nondeductivist understanding of theories provided by one particular version of the semantic conception to develop a nondeductivist analysis of scientific explanation. This is most easily accomplished by confining our attention to theoretical explanations.

Our examination of theoretical explanation parts company to a certain extent with earlier philosophical discussions. Theories possess rich structural properties,[2] and so one might expect these properties to bear a major burden in any analysis of scientific explanation. This proves not to be the case. Salmon, in his "Theoretical Explanation," exploits nothing more about theories than that they contain laws and theoretical entities. Nowhere in his discussions of explanation does Hempel exploit any of the detailed structural characteristics of theories displayed by the standard or received view of scientific theories or by later incarnations that he helped develop. And theory structure plays no important role in other analyses of explanation, such as those offered by Sylvain Bromberger or Stephen Toulmin.[3] It is a central contention of this paper that only by exploiting the structural properties of theories can we come to an adequate philosophical understanding of scientific explanation. The nondeductivist, semantic conception of theories will further enable us to exploit our understanding of the structure of theories to obtain a nondeductivist perspective on explanation. The third section of the paper will sketch just as much of the semantic conception's nondeductivist analysis as is needed to develop our nondeductivist account of theoretical explanation.

Virtually all recent philosophical discussions of scientific explanations have assumed that they are always explanations *why* or can be reduced to explanations *why*. The third section challenges that assumption and argues that explanations *how could* are more basic than explanations *why*. Once all these preliminaries have been dealt with, the heart of the paper, "Theoretical Explanation on the Semantic Conception," presents a comprehensive and somewhat detailed nondeductivist analysis of theoretical explanation based on my version of the semantic conception of theories and shows how this account is able to resolve or escape the sorts of difficulties that have scuttled earlier, deductivist analyses. One distinctive feature of the nondeductivist analysis is that it treats explanations *how could* as being more fundamental than explanations *why*.

Philosophical Analyses of Explanation

Modern discussions of scientific explanation begin with the covering law model developed by Hempel in a series of papers.[4] On this model, which has two versions, an explanation of some event (the explanandum) consists of a suitable argument wherein the explanandum follows deductively (the deductive-nomological, or D-N, model) or with high probability (the inductive-statistical, or I-S, model) from premises consisting of laws and factual conditions (the explanans). The explanans must contain at least one lawful generalization and must be true or highly confirmed; a maximal specificity condition also is imposed on the explanans. The intuitions here are fundamentally deductivist, as the D-N deductive case is the basic one and I-S explanations are a later addition that restricts statistical explanations to quasi-deductive ones in which the explanans follow from the explanandum with high probability.

A number of objections have been raised against this covering law model. First, a number of counterexamples have been advanced against the D-N model by Bromberger, Salmon, and others.[5] In the flagpole example, the D-N model allows a law of coexistence relating the angle of the sun, the height of a flagpole, and the length of its shadow to be used to explain not only the length of the shadow but also the height of the flagpole; the former seems legitimate and the latter illegitimate. This is typical of the D-N counterexamples in that it involves a law of coexistence. For reasons that will be developed below in the subsection "'Why' Explanations Reconsidered," I do not find such objections compelling and believe they can be blocked by not allowing genuine laws of coexistence to be the only laws in D-N explanations.

Second, counterexamples have been advanced by Salmon and Richard Jeffrey[6] against the I-S model. These typically involve statistical laws that are causally irrelevant. For example, the I-S model allows the law "Persons who regularly take birth control pills have a low probability of getting pregnant" and the fact that Mr. Jones regularly takes birth control pills to explain why Mr. Jones is not pregnant. And the generalization "People who wear high-heeled pumps have a high probability of wearing lipstick" and the fact that Bob went to the restaurant in drag wearing high-heeled pumps, according to the I-S model, explains why Bob was wearing lipstick. What these counterexamples point to is that Hempel never gave (nor claimed to give) an adequate characterization of what a statistical law is, and that not all statistical generalizations (including those in the examples) constitute explanatory statistical laws. While this is a serious defect in the I-S model, I believe it fails to establish that, given an adequate characterization of statistical laws, the I-S model would be unsatisfactory.[7]

Third, Hempel himself has argued [8] that D-N explanations cannot meet the requirement of being deductively valid since laws are always relative to unstated *ceteris paribus* conditions that violate deductive validity for the argument. I do not find this convincing since I see no reason why the laws cannot be stated "*Ceteris paribus,* all *A* are *B*" and why the statement "The *ceteris paribus* condition is satisfied" cannot be included among the factual conditions in the explanans. These amendments would render the argument deductively valid.

There is, however, one objection to the covering law model that I do find convincing. Salmon and Jeffrey claim that explanations are not arguments at all.[9] They argue the case as follows. The I-S model requires that the explanandum follow from the explanans with high probability. Thus on the I-S model only high-probability events can be explained statistically. But, in fact, extremely low probability events admit of statistical explanations (the authors give compelling examples). But such explanations cannot be inductively sound arguments. Hence the I-S model is defective, and statistical explanations are not arguments. The D-N model also is defective in this respect. For causal explanations are limiting cases of statistical explanations, and if statistical explanations are not arguments, neither are causal ones. So the D-N model of causal explanations also is defective. I am in basic agreement with Salmon and Jeffrey on this point. Although we often give arguments in presenting explanations (because we usually are trying to convince somebody), I do not believe that casting explanations as arguments adequately captures the structure of explanations. Thus I reject the D-N and I-S covering law models as seriously flawed. This rejection involves abandoning not only the deductive argument perspective of the D-N model but also the deductivist intuitions underlying the I-S model. If a deductivist analysis of explanation is forthcoming, it will have to be one that doesn't liken explanations to arguments.

Those who have criticized and rejected the covering law model have also tried to present alternatives. Bromberger and Toulmin present models of explanation that resemble Hempel's models in many respects but introduce new ingredients designed to avoid deficiencies in that account. At the heart of both their accounts is the idea that the scientific laws employable in explanations are highly idealized and neither directly describe nor directly apply to most actual circumstances in which explanations are needed. Bromberger calls such laws *general rules*[10] and Toulmin calls them *ideals of natural order.*[11] Bromberger maintains, because of cases such as that of the flagpole, that Hempel's D-N model provides necessary but not sufficient conditions for causal explanations; and he attempts to augment the D-N model by additional requirements that exclude these problematic cases. Central to these is the idea that explanations are required only when there are deviations from the idealized circumstances specified in the general rule. At the core of an explanation is the expansion of the general rule to an *abnormic law* that contains a finite list of all the possible deviations from the general rule and the effects of each such deviation. I find this account to be fundamentally defective for three reasons. First, it presumes that all deviations from the idealized circumstances (hence all *ceteris paribus* conditions) in principle are finitely specifiable, but this generally is not the case. Second, in many cases we do in fact know how to give explanations (e.g., using classical particle mechanics), even though we do not know how to specify all the deviations from the idealized conditions or what their effects are; hence we cannot supply the abnormic law his account requires. Third, it appears that the only reason why cases such as that of the flagpole are disallowed is that they are in accordance with the general rule—Bromberger maintains that explanations cannot be given in such cases, a point I will dispute in a moment. Thus I find Bromberger's account seriously defective. His insistence that *ceteris paribus* conditions be finitely statable seems motivated by a deductivist perspective wherein explanations are arguments. Our rejection of his analysis here casts further doubt on the viability of such a deductivist portrait.

Toulmin's account does not adhere as closely to Hempel's D-N model as does Bromberger's, but it nevertheless shares a number of Bromberger's key insights. For Toulmin, events in accordance with ideals of natural order are self-explanatory, and thus require no explanation. But no phenomena are ever in accord with the ideal (e.g., Newton's first law concerning inertia), and so to explain such deviations one must take recourse to other laws, which account for deviations from the ideal of natural order. Unlike Bromberger, Toulmin does not attempt to give a formal analysis of this procedure, and he does not commit himself to the finite specificity of all deviations from the ideal. I have two main complaints. First, the account is inadequately precise. Second, unlike both Toulmin and Bromberger, I do think that events in accordance with the idealized conditions specified in laws admit of explanations via the laws. Their explanations rest on the fact that they realize or approximate the idealized circumstances of empirically true or validated laws and behave in conformity with those laws. I agree with Toulmin and Bromberger, however, that an important aspect of explanation is the use of laws to explain events that do not meet the idealized conditions of the laws and hence are not in conformity with them. Unlike Bromberger I do not believe these deviations can be handled by anything like a finitely specified abnormic law, but I agree with Toulmin that in explaining such deviations we do have recourse to other laws or theories. How this is accomplished in theoretical explanations will be discussed later.

Salmon, in *Statistical Explanation and Statistical Relevance,* presents a much more radical response to Hempel. At the heart of his

nondeductivist treatment are the ideas (a) that statistical explanations are more basic than causal explanations, the latter being, as
mentioned above, a limiting case, and (b) that the essence of statistical explanation consists in citing factors that are statistically relevant to the explanandum event. Roughly the idea is this: Suppose
we wish to explain statistically the event B of kind A. We partition
A into a set of homogeneous classes $AC_1, \ldots, AC_n$. (The AC_i are
homogeneous if none of their various subdivisions—called place
selections—are statistically relevant to the occurrence of B.) Further, the partition must be such that:

$$P(B, AC_1) = p_1$$
$$\vdots$$
$$P(B, AC_n) = p_n$$

where $p_i \neq p_j$ $(i \neq j)$. Then to explain why $x \, \varepsilon \, A$ is B, we determine which A_k $(1 \leq k \leq n)$ x is in; the explanation of why x is
B would be "$x \, \varepsilon \, AC_k$." That is, among the A's, the statistically
relevant property responsible for x's being B is C_k. To avoid various
statistical paradoxes, two further rules are necessary. The *reference
class* rule requires one to choose the broadest homogeneous reference classes to which x belongs. The *screening-off* rule says that if
there is a choice of properties with which to make a statistically
relevant partition of a reference class, and one property screens off
the other (where D *screens off* C from B in reference class A iff
$P(B, ACD) = P(B, AD) \neq P(B, AC)$), then the screening-off
property is to be used in preference to the screened-off property.[12]
The point of this rule is to eliminate such counterexamples as
Michael Scriven's barometer case, in which two effects (barometer
and storm) of a common cause (certain atmospheric conditions) are
statistically relevant to each other and so, if allowed, would yield
spurious explanations (e.g., the barometer change caused the
storm).

There are a number of difficulties with this statistical relevance
(S-R) account. First, the homogeneity and screening-off requirements are fundamentally at odds with everyday explanations of a
practical sort, since the requirements in effect sanction only those
explanations that are expressed in the very deepest physical terms.[13]

Second, on Salmon's account causal explanations are supposed
to be limiting cases of statistical explanations—that is, ones in

which $P(B, A) = 1$. Let A = "The flagpole is α feet tall," B = "The shadow is β feet long," and C = "The sun is at angle θ over level terrain." The law involved in the flagpole case is such that $A \cdot$ *implies* B and $B \cdot C$ implies A. Thus we have

$$P(B, AC) = 1$$
$$P(B, A\ C) = 0$$

in accordance with the conditions of Salmon's analysis. But we also have

$$P(A, BC) = 1$$
$$P(A, B\ C) = 0$$

in accordance with the conditions of Salmon's analysis. Furthermore, because the probabilities of $P(B, AC) = 1 = P(A, BC)$, there is no way the screening-off rule is applicable; AC and BC are homogeneous classes,[14] and broadening the reference class will not help. Thus we have the flagpole counterexample all over again. The point here is that in the causal explanation cases a temporal asymmetry is crucial to explanatory capability: AC explains B because there is a tiny time lag between a light ray's hitting the flagpole and the resulting shadow's being cast, but BC does not explain A because that temporal asymmetry is violated. If Salmon's account cannot capture this crucial causal asymmetry in the limiting case, there is reason to doubt it can in other genuinely statistical cases.

Third, on somewhat different grounds (involving various counterexamples), Salmon has come to the conclusion that statistical relevance cannot capture the asymmetries required for explanations, and now sees causal notions as having a most fundamental role to play:

> Although we believe that the S-R model of explanation has certain virtues, we do not believe it can provide a fully adequate account of scientific explanation. In order to have any hope of achieving a satisfactory treatment of this notion, we must supplement the concept of statistical relevance with some kinds of causal considerations. . . . The relation of cause to effect has a distinct temporal asymmetry; causes come before effects, not after them. The relation of explanatory facts to the fact-to-be-explained has a similar asymmetry. In the D-N and I-S models, there is no demand that the explanatory relation be any kind of causal relation. . . . The same is true of S-R explanations; they are not

explicitly causal, and it is not clear that they embody any temporal asymmetry requirements. . . . Only by introducing causal considerations explicitly, it appears, can we impose the appropriate temporal asymmetry conditions upon our scientific explanations.[15]

To the end of imposing such conditions Salmon has been developing an account based on "continuity" and the use of "marks."[16]

I am sympathetic with Salmon's earlier view that statistical explanations are more fundamental than causal ones—though I would claim that it is indeterministic explanations that are more fundamental than deterministic ones. Further, he is absolutely correct that there is a fundamental temporal asymmetry in explanation—and I concur in the opinion that screening-off and statistical relevance conditions do not capture that asymmetry. I am, however, far less sanguine about the prospects of his "marks" and "continuity" criteria for capturing that asymmetry—in part for reasons that overlap qualms that L. Jonathan Cohen has registered.[17] More fundamentally, many scientific theories that employ discrete, not continuous, time parameters (e.g., the genetic theory of natural selection, which measures time in "generations") seem quite capable of providing explanations but do not meet Salmon's continuity requirements; thus I doubt whether that requirement will capture the essential asymmetry of explanation, except, possibly, for certain special cases. Further, I am convinced that the structure of theories revealed by the semantic conception has all that is needed to capture that asymmetry without recourse to specific analyses of causality or hypotheses about the role of "marks" therein.

The intuitions underlying Salmon's S-R model seem to me fundamentally nondeductivist, and I suspect that his replacement "marks" analysis may involve at least a limited retreat toward a more deductivist perspective. Be that as it may, if I am correct that my version of the semantic conception has the resources to provide the sort of asymmetry Salmon needs, then it follows that the problems his analyses encounter give us no reason to abandon the attempt to provide a nondeductivist analysis of theoretical explanation. Absent any suggestion about what a deductivist analysis of explanation would look like that did not liken explanations to deductive or quasi-deductive arguments, I conclude that a nondeductivist approach to analyzing theoretical explanation is most promising.

Are All Explanations Explanations
Why?

Virtually all the philosophical literature on explanation assumes tacitly, if not explicitly, that explanations are explanations *why* or responses to explanation-seeking "why" questions. While Salmon and others assume this, to my knowledge only Hempel has attempted to defend the view; and then only briefly:

> A scientific explanation may be regarded as an answer to a why question, such as: "Why do the planets move in elliptical orbits with the sun at one focus?" "Why does the moon look much larger when it is near the horizon than when it is high in the sky?" "Why did the television apparatus on Ranger VI fail?" "Why are children of blue-eyed parents always blue-eyed?" "Why did Hitler go to war against Russia?" There are other modes of formulating what we will call *explanation-seeking* questions: We might ask what caused the failure of the television apparatus on Ranger VI, or what led Hitler to his fateful decision. *But a why-question always provides an adequate, if perhaps sometimes awkward, standard phrasing.*[18]

Is it really the case that every explanation-seeking question is equivalent to a "why" question—hence that all explanations are equivalent to answers to such questions?

Consider a partial list of possible erotetic descriptions of explanations:

(1)   The anthropologist explained *who* slept in the village longhouse.

(2)   The technician explained *what* the significance of the peaks on the oscilloscope pattern was.

(3)   The biologist explained *where* to look for sea urchins.

(4)   The man explained *why* the event *did* happen.

(5)   The man explained *why* the event *was possible*, though improbable.

(6)   The biologist explained *what* hemoglobin does.

(7)   The biologist explained *when* the grunion run.

(8)   The physician explained *how* the multiple birth *did* occur.

(9)   The physician explained *how* the multiple birth *could have* [might have] occurred.

(10) The economist explained *which* product's production *would* maximize marginal cost.

Some of these questions (e.g., the second) are such that there is a corresponding "why" question that could have been asked, and such that an answer to the indicated question would contain the ingredients of an answer to the corresponding "why" question. But erotetic logic today is sufficiently advanced[19] that we know that different sorts of questions have different presuppositions and that what counts as an answer to a question is dependent upon the answer's standing in the appropriate logical or empirical relationship to the question's presupposition. In particular, answers to "who," "where," "what," "whether," etc., questions generally entail their presuppositions whereas the answers to "why" and "how" questions generally do not.[20] Since "why," "how," "who," "where," "what," etc., questions have different presuppositions, it is not clear that the answer to an explanation-seeking non-"why" question automatically qualifies as an answer to any corresponding "why" question. So, for example, some answers to the "who" question in (1) above may answer the corresponding "why" question while others may not. Hempel's thesis thus seems suspect at best—and certainly in need of more sustained defense than he has offered.

My concern is with theoretical explanation, for which the explanation of isolated events is not central; thus it may seem that such qualms over the "why" explanation equivalences of, for example, questions (1), (2), (6), (7), and (10) may be beside the point. In these cases it may be (though I remain unconvinced). But, I would maintain, there are types of theories that afford a variety of explanations *how* but are incapable of providing explanations *why*—in which case the point of my objection is fundamental. There is a variety of known classes of systems that, in the ideal at least, are appropriately characterized as finite Markov processes or chains. For the sake of quick illustration suppose that we have such a class of systems, characterized by four states, which is governed by the following state-transition matrix. This matrix, of course, determines a probabilistic state-transition branching tree—in which all paths are equiprobable. Suppose that at t we know what state s some particular such system is in and that at $t + 34$ the system is in state s'. Since the number of paths from s to s' is exceedingly large, and since each path is equiprobable, I claim this Markov theory is incapable of explaining on the basis of the system's being

State at $t + 1$

	1	2	3	4
1	1/4	1/4	1/4	1/4
2	1/4	1/4	1/4	1/4
3	1/4	1/4	1/4	1/4
4	1/4	1/4	1/4	1/4

(rows labeled "State at t")

in state s at time t *why* the system is in state s' at time $t + 34$. I do claim, however, that it could explain *how* the system in state s at t *could have* come to be in state s' at $t + 34$—namely, by showing that there is a *possible* state-transition path from s to s' in thirty-four state-transitions. Indeed, there will be an inordinately large number of such "how could" explanations that are equiprobable. If I had in addition *full* knowledge of each of the state transitions at intermediate times from t to $t + 34$, I would be able to explain via the above matrix (law) *how did* a system in state s at t *did* come to be in state s' at $t + 34$; but I would not be able to explain *why* the system in state s at t was in state s' at $t + 34$—for I cannot explain *why* the various state transitions that occurred at intermediate times did occur (as opposed to other possible and equiprobable ones).

The situation would not be significantly different if one had larger matrices or matrices with radically unequal probabilities in the various cells; indeed, even the existence of absorbing states would not appreciably alter the situation. Crudely overstated, the situation is this: If a theory generally allows at least one theoretically possible state-transition path from state s at t to state s' at t' ($>t$), then a "how could" explanation for the system's being in state s' at t', given that the system previously was in state s at t, is possible. If one also knows the state-transition history between s and s', one can give an explanation of how the system, previously in s at t *did* come to be in s' at t'; but only if one can show, via the theory and its laws, that a system in state s at t *had* to go into state s' at t' ($>t$) can one give an explanation of *why* that system in state s at t is (or was) in state s' at t'. In short, an explanation *why* is an explanation *how could* coupled with a uniqueness claim, namely, "and that's the *only* thing that could have happened."

Not only do I deny that all explanation-seeking questions are reducible to equivalent explanation-seeking "why" questions but I am convinced that explanation-seeking "how could" questions are explanatorily more basic. Thus I believe Hempel and also Salmon and others are mistaken in their assimilation of explanation to explanation *why*. More important, I believe that full appreciation of this point leads to our treating indeterministic (as opposed to deterministic) theoretical explanation as the basic case—and that our doing so ultimately displays in part why Salmon's recent retreat to causality, continuity, and marking conditions misses the mark.

My conclusion is that explanations *how could*, not explanations *why*, are the basis for an adequate philosophical analysis of theoretical explanation. Further, I believe that the semantic conception of theories has the resources to provide an adequate understanding of such explanations—and I now turn to the task of showing that this is so.

The Semantic Conception of Theories

By now many of my cards are out of my sleeve. I have outlined what I perceive to be a number of the strengths and weaknesses of recent and classical treatments of scientific explanation, and have indicated the alternative general positions I would take on these issues. If it is not already abundantly clear that my stands are strongly reinforced by my version of the semantic conception of theories, it soon will be. Accordingly, the development of the non-deductivist, semantic conception deserves a summary account.[21]

On the semantic conception, theories are concerned with specifying the behaviors of systems of entities, and construe the behavior as changes in a specified finite set of parameters characteristic of these entities. At a given time the simultaneous values of these parameters determine the *state* of the system. The *intended scope* of a theory is some natural class of physically possible—which is to say, causally possible—systems (e.g., the class of all causally possible mechanical systems composed of a finite number of interacting bodies). It is the job of a theory to characterize that class by indicating all and only those time-directed sequences of states which correspond to behaviors of possible systems within its intended scope.[22]

What is the structure of a theory, and how does that structure enable it to do the job just described? In essence a theory is a general model of the behavior of the systems within its scope. The model is a relational system whose domain is the set of all logically possible state occurrences in the systems within the theory's scope. The model's relations determine time-directed sequences of state occurrences which correspond to the behavior of possible systems within its intended scope and indicate which changes of state are physically possible. These sequencing relations are the *laws* of the theory. A variety of sorts of laws are possible which differ in the ways they determine possible sequences of state occurrences. The various sorts of laws will be discussed later.

A scientific theory is an iconic model. However, as is the case generally with such models, the correspondence determined by a theory need not be one of identity. For in specifying possible changes in state, the theory tacitly assumes that the only factors influencing the behavior of a system are those that show up as state parameters in the theory, whereas in fact this behavior—and hence the actual values of these parameters—often is influenced by outside factors that do not show up as distinct parameters of the theory. Thus the nature of the correspondence is as follows: *The sequences of states determined by the theory indicate what the behaviors of the possible systems within the theory's scope would be were it the case that only the factors modeled by the parameters of the theory exerted a nonnegligible influence on those behaviors.* That is, the theory characterizes the possible behaviors of systems under idealized circumstances wherein the values of the parameters do not depend on any outside influences, and thus relates counterfactually to many actual systems within its intended scope.

According to this semantic conception of theories, then, scientific theories are relational systems functioning as iconic models that characterize all the possible changes of state that the systems within their scopes could undergo in idealized circumstances. And the theory will be *empirically true* if and only if the class of possible sequences of state occurrences determined by the theory is identical with the possible behaviors of systems within its intended scope under idealized conditions. Whenever a system within the theory's intended scope meets the idealized conditions, the theory can predict the subsequent behavior of the system (the preciseness of the prediction depending on what sorts of laws the theory has, as indicated above). When the idealized conditions are not met, the theory can predict the behavior of the system *if* used in conjunction with a suitable auxiliary theory or law; only in some cases are suitable auxiliary laws or theories known.

Theories are not always propounded as candidates for being empirically true. Rather they may be propounded either heuristically or as simplifications or when known to be false. When theories are put forth as such *conceptual devices*, deep questions arise as to their potential for providing explanations.[23] In the present paper I will ignore such complications, confining my attention to the more basic question how empirically true theories can provide, or function in, explanations.

Theoretical Explanation on the Semantic Conception

I now turn to the implications of the semantic conception of theories, as informed by the foregoing discussion, for the nondeductivist analysis of theoretical explanation. Although it is common to formulate theories using multiple laws (e.g., Newton's three laws), on the semantic conception theories with multiple laws are always equivalent to theories with single laws.[24] So we may assume here that theories have only one law and investigate the explanatory capabilities of the main sorts of laws. In characterizing these laws I need to introduce two notational conventions. First, depending upon the theory, time may be construed as having the order properties of the natural numbers (an ω sequence), of the integers (an $\omega^* + \omega$ sequence), or of the real numbers (a λ sequence); any simple ordering of times having the order properties of an ω, $\omega^* + \omega$, or λ order type will be known as an α-*time sequence*. Second, the defining parameters $p_1, \ldots, p_n$ of a theory are variables; let $p_i(t)$ be the value of parameter p_i at time t. Then the *state s* of a system at time t is $<p_1(t), \ldots, p_n(t)>$. (Where needed, $q_i(t)$ and $r_i(t)$—with or without primes or subscripts—will be construed analogously.) Since the same state s may occur at more than one time (i.e., in distinct state occurrences), we will use "$s(t)$" to designate the occurrence of state s at time t. D will be the set of all logically possible state occurrences for a theory.

We now are in a position to analyze the sorts of explanations that different kinds of laws can provide via theories construed in accordance with the semantic conception.

"How Could" Explanations with Statistical Laws of Succession

I begin with a consideration of statistical laws of succession.

Definition: Let t, t' be variables ranging over an α-time sequence. Then a *statistical law of succession* is a relation $R[s(t), s'(t'), p]$ meeting the following conditions for each t and t' such that $t < t'$:

$$<s(t), (s'(t'), p> \ \varepsilon \ R \ \text{iff} \ P[s'(t'), s(t)] \ = \ p,$$

where P satisfies the axioms for the conditional probability operator, and

$$\text{for each } s(t), \sum_{s'(t') \ \varepsilon \ D} P[s'(t'), s(t)] \ = \ 1.$$

The key feature of such laws is that for a given current state, the given law allows that the system may enter a number of different subsequent states and assigns conditional probabilities that these various subsequent states will be entered. The Mendelian laws of genetics and finite Markov processes are paradigmatic examples.

Suppose we are concerned with the way molecules of a gas behave when put in a compartment containing a thin, semipermeable membrane that divides the compartment into equal-sized cells. Suppose further that we know there are $2n$ molecules of gas (for some n), and that we have the following empirically true theory T about such systems: The possible states are $<j, k>$, where j is the number of molecules in cell 1 and k the number in cell 2, and the state-transition law governing the behavior of this system is characterized by:

$$
\begin{aligned}
k &= 2n - j \\
P(j-1,\ j) &= j/2n \\
P(j+1,\ j) &= (2n-j)/2n \\
P(m,\ j) &= 0 \text{ otherwise.}
\end{aligned}
$$

(On this theory it is assumed that exactly one molecule crosses the membrane at each time step.)[25]

Suppose I walk into my lab, examine cell 2 and find it empty, then count $2n$ molecules in cell 1—that is, the compartment is in state $<2n, 0>$. Perplexed, I seek an explanation. I ask the lab assistant what he's done recently to the apparatus, and he replies that two hours ago he cleaned the apparatus, then introduced n molecules of gas into each cell—and that he hasn't touched a thing since.

Drawing on my theory (above) I now am able to explain how state $<2n, 0>$ could have happened. I use the theory to draw a branching tree, which begins as shown below (the p_i that label branches are probabilities, not necessarily distinct).

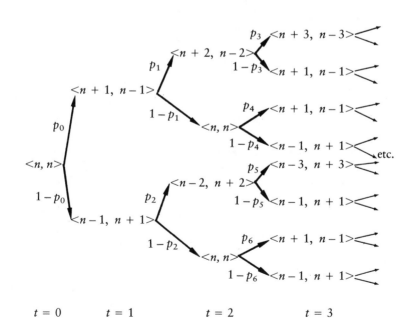

I calculate the number of time steps that elapsed from the time the apparatus was filled until I observed state $<2n, 0>$ at t'. I then determine that, yes, state $<2n, 0>$ is on several of the nodes for the t' column of the tree, and then see that (since $t' = t + 2$ hours $\gg 2n + t$) there are multiple paths down the tree from the root to each of those nodes. I now have an explanation of how the apparatus could be in state $<2n, 0>$ at t'—namely, there is one (or more) possible path(s) from known state $<n, n>$ at $t = 0$ to observed state $<2n, 0>$ at time t'.

The foregoing example is a case in which the actual situation approximates the fiction that only the defining parameters of the system affect the system's state-transition behavior—that is, that the system is isolated from outside influences. Suppose, however, that when I examined the compartment contents I found not state $<2n, 0>$ but instead state $<2n - 16, 0>$. The above theory does not allow this as even a possible state, so by itself T affords no explanation. Knowing this I suspect that the apparatus does not meet the isolation conditions of the theory. On careful inspection I discover that there are several microscopic breaks in the welded corners of cell 1, large enough for a gas molecule to escape. Then I experimentally determine the likelihood that a molecule will escape over a time step. I combine this new "local law" with the original theory (relativizing its equations to a variable number of molecules in the system) to get a new special-case theory that covers the situation. I construct a revised state-transition tree, determine the existence of one or more paths from known prior state $<n, n>$ at $t = 0$ to state $<2n - 16, 0>$ at t', and thereby explain how the system could be in state $<2n - 16, 0>$ at t'.

Some comparison of these two cases is in order. In the second case the isolation conditions presupposed by our basic theory T were not met. But once we found what the circumstances violating the isolation condition were, established a local law governing them, and combined this with the basic theory T, we produced a new, local theory T'. And T' imposes its own conditions of isolation from outside influence; relative to these our apparatus was isolated from outside influence. This makes it clear that theories are capable of providing "how could" (and other) explanations *only* of systems that are isolated from outside influences not taken into explicit account by the theories—and that in actual practice we sometimes can use local laws in conjunction with quite general theories to

produce explanations of systems that fail to meet the isolation conditions of the general theory. In either case our explanation of the singular event is a "how could" explanation of such events in *all* systems meeting the isolation conditions of the theory providing the explanation, and so accounts for entire classes of phenomena.

It is noteworthy that this account of statistical "how could" explanations does not make direct recourse to any notions such as "statistical relevance," requirements of "maximal specificity," or causal notions. Some comments on this are in order. First, consider the indeterministic nonstatistical analogue of statistical "how could" explanations—that is, consider a theory that indicates possible state-transitions without assigning any probabilities to the state-transitions. Examples of such theories include certain grammatical theories with optional transformation rules, as well as a number of anthropological theories about social structure, such as that describing certain tribes in which some young boys are selected to become berdache shamans (transvestite medicine men or priests) while most are selected to become warriors. From such theories one can explain how certain phenomena could be—even though one cannot assign probabilities to the explanandum states. This strongly suggests that statistical considerations are not essential to "how could" explanations, but rather greatly enhance the predictive capabilities of a theory.

Second, in the case of statistical theories it might seem that statistical relevance conditions are necessary to avoid conjunctive forks, the introduction of superfluous parameters, etc. I do not agree: the requirement that the theory be empirically true is strong enough. If a theory meets the empirical truth requirements imposed above (which are quite strong), it is virtually certain that none of its defining parameters are superfluous or artificially included. Statistical relevance considerations may have a legitimate place in the confirmation or testing theory utilized to determine whether a theory is empirically true, but that does not require their inclusion in any analysis of theoretical explanation.

Third, statistical laws of succession have built-in the sort of temporal asymmetry that is required for explanation, without explicit recourse to any causal notions.[26] Moreover, it is unclear whether any causal notions—even statistical ones—truly are built into statistical laws of succession. It may be that on some statistical theories of causation, statistical laws of succession are causal—but if so, it is not obvious to me that this is the sense of "causal" required to

give explanations *why*. Nevertheless we will see below that there are rather peculiar circumstances in which statistical laws of succession do provide explanations *why*. In my opinion the discussion there lays the basis for doubting that there is any simple connection between causation and the ability to provide *why* explanations.

"Why" Explanations with Deterministic Laws of Succession

"Why" explanations typically (but not inevitably) are supplied by theories with deterministic laws of succession. These laws are characterized as follows.

Definition: Let t, t' be variables ranging over an α-time sequence. Then a deterministic law of succession is a relation $R[s(t), s'(t')]$ meeting the following condition for every t and t' such that $t < t'$:

if $<s(t), s'(t')>$ ε R *and* $<s(t), s''(t')>$ ε R, *then* $s' = s''$.

The equations of motion for classical particle mechanics specify a paradigmatic deterministic law of succession. Intuitively, the characteristic feature of such laws is that the current state of a system determines unique subsequent states. If we view laws of succession as specifying state-transition trees,[27] the main difference between deterministic and indeterministic laws is that, whereas the latter allow multiple paths to branch off from each node, the former allow only one state-transition path from each node. This has the consequences that indeterministic theories may have multiple paths from a prior state at t to the explanandum state at t' and that it is possible that the system might have been in some other state at t'. Both possibilities are precluded by deterministic laws of succession. Thus if we know that a system was in state $s(t)$, and we want to explain its being in state $s'(t')$, we are able to explain *how* the system *could be* in state $s'(t')$: we have an empirically true theory with a deterministic law of succession that applies to the system; the system meets the law's isolation conditions; and the theory specifies a state-transition path from $s(t)$ to $s'(t')$. But the theory tells us more: it tells us that this is the *only* state that the system could be in at t' and that there is a *unique* path whereby it came to be in $s'(t')$—given that it previously was in $S(t)$. In short, the theory tells us not just how the system could be in $s'(t')$ but that

that's how it had to be. It provides an explanation of *why* it is in $s'(t')$.

This suggests that an explanation *why* is an explanation *how could* such that there is a unique state the system could be in at the time of the explanandum event or state and a unique path to that state from a known prior state. We will evaluate this suggestion below.

We can now say something more about the relations between explanations afforded by statistical laws of succession and explanations *why*. First, consider our examples from the previous subsection. Suppose that we augmented our explanation of how the apparatus could be in the explanandum state with detailed knowledge of each state-transition between t and t'. In this case we would have an explanation of *how* the system *did* come to be in the explanandum state; but we would *not* have an explanation of *why* it came to be in that state at t'—since it could come to be in that state by a different path and it could have been in a different state at time t'. There are, however, circumstances in which a theory with a statistical law of succession can provide explanations *why*. Consider, for example, the following portion of the state-transition tree defined by a statistical law of succession.

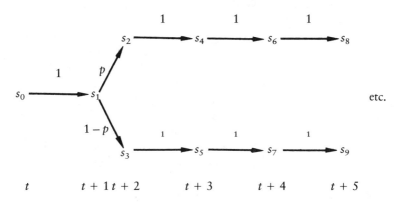

Such a law could only explain *how* a system in state s_0 at t *could* be in state s_8 at $t + 5$; it could not explain *why* it was. But it could *explain why* the system in state s_2 at $t + 2$ was in s_8 at $t + 3$.

Consider now a system such as a pinball machine, which is governed by a finite Markov process with a terminal absorbing

state where all paths lead to the terminal state (e.g., the ball leaves the playing area through a single slot). Let s_I be the initial state and s_A be the terminal absorbing state. To visualize this, take a simplified pinball system with three states whose state-transition matrix and state-transition tree are shown below.

State at $t + 1$

		s_I	s_1	s_A
	s_I	0	1/2	1/2
	s_1	0	1/2	1/2
	s_A	0	0	1

State at t

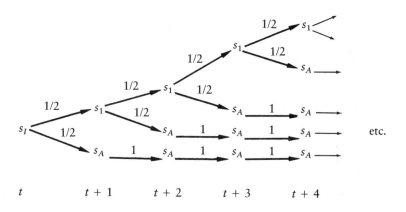

| t | $t + 1$ | $t + 2$ | $t + 3$ | $t + 4$ |

The characteristic features of this tree are that multiple state-transition paths are possible, but that all paths eventually lead from s_I to s_A. Can the correct theory for this sort of system provide explanations *why*? Initially I thought not, my intuition having been that the unique-terminus and unique-path requirements suggested above were essential for explanations *why*. However, a number of people at the workshop on which this volume is based, as well as a number of my better graduate students, have pressed intuitions that in cases such as this the theory does enable one to explain why the system, initially in state s_I, ended up in s_A. For a long time I resisted that intuition. I now believe that the basis of my resistance was that I was construing the request for an explanation *why* as a request to

explain why the system initially (at t) in s_I, *now is (or at time t'* was) in state s_A, and the unique-path as well as the unique-terminus requirement is needed to do this. (In the simplified example above, which path is taken uniquely determines the time at which the system goes into s_A.) But the contrary intuition seems to construe the request for an explanation *why* as a request to explain why the system *eventually* come to be in state s_A; and this time-unspecified explanation does not seem to require the unique-path condition, though it does require the unique-terminus condition. This suggests that there are two sorts of requests for explanations *why*: explanations of why something had to happen and explanations of why something had to happen when it did. Thus, we could distinguish two sorts of explanations *why*: *Strong explanations why* explain why a system in state s at t had to be in state s' at time t', and thus are explanations *how could* that meet the originally proposed unique-path and unique-terminus requirements. *Weak explanations why* explain why a system in state s at t *eventually* had to end up in state s' (but not why it did so at a particular time), and thus are explanations *how could* that meet the unique-terminus but not the unique-path requirement.

Is this suggestion viable? To see, let us reconsider our simplified pinball machine example. It clearly does not allow a strong explanation *why* since there is not a unique path to state s_A. But it also does not meet the weak explanation *why* requirement since it is not the case that the system eventually *has* to end up in state s_A—for at every time t' there is some finite probability $(1/2)^{(t'-t)}$ that it will be in s_1 [despite the fact that $\underset{(t'-t) \to \infty}{\text{Limit}} (1/2)^{(t'-t)} = 0$]. By contrast, a system governed by a statistical law of succession with the following tree portion would enable one to give a weak explanation of why

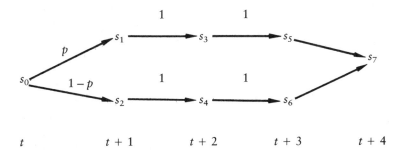

the system in s_1 (or $s_1 - s_6$) had to be in state s_7 at time $t + 4$, but since it does not meet the unique-path condition it does not afford us a strong explanation of why the system had to be in s_7 at $t + 4$. Nevertheless it does explain why the system in s_0 at t had to be in state s_7 at $t + 4$. This indicates that the unique-path requirement of strong explanation *why* is not necessary—it is a condition met only by deterministic explanations of why an event had to happen at t', but indeterministic explanations of why an event had to happen at t' are also possible. What is essential for explanations of why an event happened at time t' is that there be only one state the system could be in at time t', not how many paths lead there in the same time. What counts in explaining why Caesar got to Rome when he did is not that all paths led to Rome, but that all paths led to Rome and were of the same length.

We conclude then that our proposed distinction between weak and strong explanations *why* is unsatisfactory. Rather, in order to explain why a system in state s at t was in state s' at t', the applicable theory must be such that s' is the only state the system could be in at t'. Deterministic theories are in addition such that there will be a unique path whereby the system in s at t came to be in state s' at t'; but such unique paths are not essential to explanations *why*. Our discussion thus indicates how very subtle and complex are the relations between deterministic laws, indeterministic laws, and the ability to provide explanations *why*. It should be noted that explanations *why* automatically provide explanations *how could*, but only explanations *why* based on deterministic laws of succession automatically provide explanations *how did*.

I have not bothered to illustrate the discussion in this section with even simplified versions of deterministic scientific theories since readers are already familiar with them. Absent such examples, I should stress that, as with indeterministic theories, deterministic theories can provide explanations only for systems that meet the isolation conditions for the theory. But in cases of nonisolation, given suitable auxiliary laws or theories, we frequently can produce special-purpose theories for which our explanandum systems do meet the isolation conditions. Thus, for example, in classical particle mechanics we can produce special-case theories concerning terminal velocity or frictional inclined planes that are applicable to some sky-diving and loading-dock situations because their isolation conditions are met.

Finally, I note in passing that it is allowed that the defining
parameters for states in a theory with a deterministic law of succession may be probability distribution functions; a paradigmatic example is quantum theory. In such cases one can explain *why* a
system was in a particular quantum state at time t', but one can
only explain (statistically) *how* a particle of the system *could have*
a particular observed position or momentum.

Explanations with Laws of Quasi-Succession

Various sorts of laws have been identified in the literature in
addition to laws of succession. Among the more prominent are laws
of interaction and teleological and functional laws.[28] It can be
shown that these are special cases (as are laws of succession) of a
more basic sort of theoretical law—*laws of quasi-succession.*[29]

Definition: Let t and t' be variables ranging over an α-time
sequence, let $P_1, \ldots, P_n$ be the basic parameters of the theory,
and let $i_1, \ldots, i_k$ ($k \leq n$) be a nonredundant listing of
parameters among $p_1, \ldots, p_n$; for simplicity, suppose that
$s(t) = <i_1(t), \ldots, i_k(t), p_{k+1}(t), \ldots, p_n(t)>$. Then $R[s(t),
s'(t')]$ is a *deterministic law of quasi-succession* iff the follow-
ing condition is met for any t, t' such that $t < t'$ and
$p_{k+1}, \ldots, p_n$ remain unchanged over the time interval $[t, t']$:

if $<s(t), s'(t')> \ \varepsilon \ R$ and $<s(t), s''(t') \ \varepsilon \ R$, then $s' = s''$.

Definition: Let t and t' be variables ranging over an α-time
sequence, let $P_1, \ldots, P_n$ be the basic parameters of the theory,
and let $i_1, \ldots, i_k$ ($k \leq n$) *be a non-redundant listing of the
parameters among* $p_1, \ldots, p_n$; for simplicity, suppose that
$s(t) = <i_1(t), \ldots, i_k(t), p_{k+1}(t), \ldots, p_n(t)>$. Then a *statistical law of quasi-succession* is a relation $R[s(t), s'(t'), p]$ meeting the following condition for each t and t' such that $t < t'$
and $p_{k+1}, \ldots, p_n$ remain unchanged over the time interval
$[t, t']$:

$<s(t), s'(t'), p> \ \varepsilon \ R$ iff $P[s'(t'), s(t)] = p$

where P satisfies the axioms for the conditional probability
operator, and for each $s(t)$,

$$\sum_{s'(t') \,\varepsilon\, D} P[s'(t'), \; s(t)] \;\; = \;\; 1.$$

Teleological and functional laws are just laws of quasi-succession on which goal states have been imposed.[30] It should be noted that when $k = n$, we have deterministic and statistical laws of succession as degenerate special cases of laws of quasi-succession. To simplify discussion, in the above definitions let us call $s(t)$ the *complete state* of the system at t, $<i_1(t), \ldots, i_k(t)>$ the *internal substate* of the system at t, and $<p_{k+1}(t), \ldots, p_n(t)>$ the *external substate* of the system at t. Thus what is characteristic of systems with laws of quasi-succession is that subsequent internal substates, but not subsequent external substates, are a function of the prior complete states of the system.

The law governing a particular model of an electronic digital computer is a deterministic law of quasi-succession, as are the laws governing a particular class of servomechanisms, and the law governing a Wheatstone bridge. Measurement apparatus typically is governed by a deterministic or statistical law of quasi-succession.

One of the most interesting examples of a theory with a law of quasi-succession is the genetic theory of natural selection. On this theory the internal states are genotypic makeups of an interbreeding biological population. The basic law of the theory involves various laws of chromosomal recombination during meiosis and recombination—which determine the makeup of the next generation of the population (at $t + 1$) as a function of not only the population distribution at t (the internal substate) but also the birth/death/reproduction ratios of the various genotypes (these ratios being the external substate parameters). The theory and knowledge of just one prior complete state at t provides absolutely no means for predicting subsequent external substates beyond $t + 1$.[31]

This latter fact complicates the giving of explanations. Take a simple special case: Suppose I know the genotypic makeup of a population at time t, have monitored the birth/death-reproduction ratios of the various genotypes from t to $t + 3$ (measured in generations), and have determined that genotype G is becoming extinct (i.e., at t it was α proportion of the population and at t *it is* $\alpha/100$ of the population). From the data I have I can determine how it is *possible* that the population is in internal substate

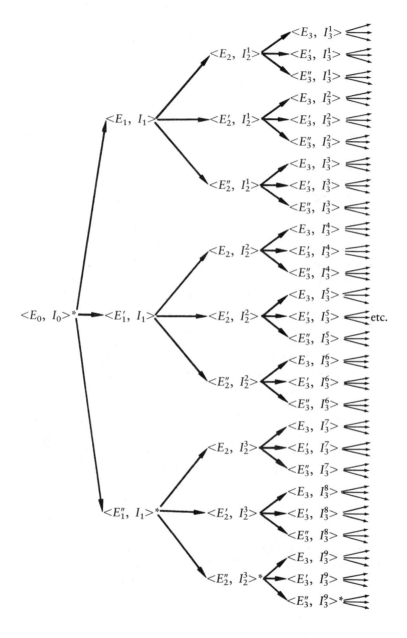

$s'(t')$ such that G has $\alpha/100$ proportionate representation in the population; but the way I do so is rather different from the way I proceed in the case of laws of succession. To see this, consider the following simplified tree representation of the theory, in which E and I represent the external and internal substates of the complete state $<E, I>$ at t, asterisks indicate the state actually entered at each time, and time is measured in generations.

Note that at any time, given the complete state, the theory enables us to predict a unique I for the next generation; but it does not enable us to predict the next or subsequent complete states. However, since we know the external substates for each time, we can in fact predict, generation by generation, subsequent internal substates and, combining that with our knowledge of the external substates, determine what the subsequent complete states will be. Indeed, the theory together with prior knowledge of the external substates at $t + 1$, $t + 2$, and $t + 3$ yields a unique path from $<E_0, I_0>$ to $<E_3'', I_3^9>$; given those external substates, $<E_3'', I_3^9>$ is the unique terminus allowed by the theory, and so we explain how the system in $<E_0, I_0>$ at t could be in complete state $<E_3'', I_3^9>$ at $t + 3$. Further, we can give an explanation of how the system in complete state $<E_0, I_0>$ at t could be in internal substate I_3^9 at $t + 3$—namely, there is a path (indeed there are three of them) from $<E_0, I_0>$ at t to I_3^9 at $t + 3$. Thus if $\alpha/100$ is G's representation in I_3^9, we can explain how it could be that G's representation has dropped so drastically as to risk extinction.

"How did" and "why" explanations are more problematic in this case. Given knowledge of the external substates the system was in at $t + 1$ and $t + 2$, we know how the system in $<E_0, I_0>$ at t came to be in I_3^9 at $t + 3$. And it seems plausible to say that we can explain how the system did come to be in I_3^9 at $t + 3$: "The system was in $<E_0, I_0>$ at t; E_1 and E_2 occurred; and so the system went into I_3^9 at $t + 3$." In the case of G's extinction, the explanation would be: "At t, G is α proportion of the population, but its reproduction rates at t, $t + 1$, and $t + 2$ were relatively low compared to others in the population. So, according to the laws of natural selection theory, G was only $\alpha/100$ proportion of the population at $t + 3$." But we would want to say that we could not explain how the system in $<E_0, I_0>$ at t did come to be in $<E_3'', I_3^9>$ at $t + 3$; for the explanation would have to consist of the "how did" explanation of its being in I_3^9 at $t + 3$ coupled with a reference to the fact that the external substate at $t + 3$ was E_3, and it seems illegitimate to include part of the explanandum in the explanans. This situation brings out the fact that in "how did" explanations only knowledge of prior states and the theory may figure in the explanans.

Are any "why" explanations possible in this case? Consider G's nearing extinction. The birth/death/reproduction ratios essentially are a summary of the effects of environmental, ecological, and other factors that affect the reproductive success of a genotype. Our path tree indicates that, given birth/death/reproduction ratios (that is, external substates) E_1'' and E_2'' at $t + 1$ and $t + 2$, there is a unique internal substate I_3^9 that the system can be in at $t + 3$. Thus the theory augmented by knowledge of E_1 and E_2 enables us to explain why the system in $<E_0, I_0>$ at t was in internal substate I_3^9 at $t + 3$. And since G having $\alpha/100$ representation is part of I_3^9 we can explain why it has gone into near extinction since t: "The population was in complete state $<E_0, I_0>$ at t; the various genotype ratios at times $t + 1$ and $t + 2$ were E_1'' and E_2''; and given the laws of natural selection theory, the system had to be in internal substate I_3^9 at $t + 3$; under I_3^9, G's representation is $\alpha/100$." Note that the explanation *why* here requires our taking into account the reproduction ratios of all genotypes of the population, not just those of G. Theories with deterministic laws of quasi-succession, together with knowledge of the external substates at t' ($t < t' < t''$), can provide explanations of why a system in complete state s at t is in internal substate i at t'' just in case there is a unique substate i the system could be in at t'' given the external substate history between t and t''.[32]

Compare the previous case with the following. Our theory is natural selection theory, but instead of knowledge of the intermediate external substates, we have an empirically true ecological theory with a deterministic law of succession which enables us to predict *inter alia* the birth/death/reproduction ratios for a population. Then we can combine this theory with natural selection theory to produce a new evolutionary theory that will have a deterministic law of succession. And, of course, it could explain (in a deeper or fuller fashion) why G was undergoing extinction. Lacking such a theory of the external substate transitions, we can still explain why the system came into internal substate i by resort to detailed, specific information about the external substate history. In the former case, in which we had the ecological theory as well, we used that theory to impose boundary conditions on the deterministic law of quasi-succession to make natural selection theory act as if it were governed by a deterministic law of succession. In the latter case, we used knowledge of the external substate history to impose local boundary conditions (for a short time span and for a specific population) to essentially the same explanatory effect. Note that in this case, while we can explain why the system is in internal substate i, we cannot explain why it is in complete state s', for that would require part of the explanandum to figure crucially in the explanans. In our combined ecological and natural selection theory, however, we could explain why the system was in complete state s' at t' given that it was in s at t; for that theory can predict subsequent birth/death/reproduction ratios.

The same process of using another theory or external substate histories to impose boundary conditions is applicable to theories with statistical laws of quasi-succession. Thus, drawing from earlier results, we conclude that when the statistical law of quasi-succession and boundary conditions require the explanandum state to be the unique possible outcome, an explanation *why* is possible. We conclude, therefore, that an empirically true theory with laws of quasi-succession, augmented with boundary conditions specifying an external substate history over (t, t'), can provide an explanation of why a system in complete state s at t is in internal substate i at t', just in case there is a unique internal substate i at t' allowed by the law and the boundary conditions.

"Why" Explanations Reconsidered

In my discussions of statistical explanations *why*, I tacitly maintained that theories with statistical laws can give theoretical explanations of why a system in state s at t is in state s' (or i) at t' only when the specified unique-terminus conditions were met. In science, however, we frequently encounter statistical explanations *why* in which a unique terminus does not obtain. For example, in the 1950s the American Cancer Society ran a study whose conclusion was that smoking causes cancer. From the statistical generalizations produced by that study one can give statistical explanations of why Jones developed cancer (because he smokes three packs a day), even though not everybody who smokes three packs a day gets cancer.[33] Since there is no unique terminus in such cases, does this show that our requirements are too stringent?

There are a number of important differences between the cancer case and theoretical explanations *why*. First, in the cancer case we do not have an empirically true theory with a theoretical law. Rather, we have various empirical statistics about the relative incidence of cancer among smokers and nonsmokers which are not equivalently recast as a theory.[34] Thus it is not a case of theoretical explanation at all. Second, the notion of causality involved in such studies is a statistical one that essentially measures causal influence,[35] whereas our account is concerned with "complete causes." Third, our account is concerned with explaining why events happen *at the time they do* and why they happen instead of some alternative event. The cancer study does neither; it does not provide explanations of why an individual got cancer when he did, or why he got it as opposed to not getting it. Fourth, what the cancer studies allow is explanations of why it was more likely Jones would get cancer than some other people. Thus, at best it enables us to explain how he could get cancer and predict the likelihood of his getting it. If (contrary to our first point) the study had resulted in an empirically true theory about cancer (with a statistical law), it would enable precisely the sorts of "how could" explanations that our account already allows but would not yield an explanation *why*.

Collectively these considerations point out that "why" explanations based on statistical notions of causality constitute no challenge to our analysis of theoretical explanations *why*. The discussion further points out how important it is in philosophical discussions of explanation to be clear and precise as to what explanation requests the analysis is concerned with: "why did," "why did . . . at *t*," "why is it likely that," "why is it possible that" are different explanation requests, and their answers require different analyses. Failure to specify the explanation request precisely is commonplace in the explanation literature and at times has hindered our understanding of scientific explanation.[36] Our discussion here also lays the basis for casting further doubt on the claim that all explanation requests equivalently can be recast as requests for explanations *why*; for it reveals that not all requests for explanations *why* are equivalent.

Our discussion of theoretical explanations also lays a basis for further assessment of treatments of explanation by Hempel, Salmon, and others. Such discussions typically employ laws of the forms:

$$\text{If } C_1, \ldots, C_n \text{ then } E$$

(*)   or

$$P(E, C_1, \ldots, C_n) = r$$

These laws (or, as I would prefer, "law statements") are *highly unrepresentative* of the sorts of theories science typically produces. Other than theories used in the design of measuring apparatus (surely "local" as opposed to "global" theories), scientific theories tend to characterize their phenomena as "closed" rather than "open" systems—which is to say that they are theories with laws of succession (or coesixtence—see the next subsection below). But the philosophical discussions of explanation largely ignore theories, looking instead at laws of the forms (*) above.

These essentially describe transducers from an input/output perspective, and amount to laws of quasi-succession with degenerate internal substates. It is natural to construe such laws as (deterministically or statistically) causal, and hence as affording explanations *why*. Here it is important to remember, and take seriously, my opening quotation from Salmon about the relative scientific importance of theoretical explanations of *general* classes of events and the relative unimportance of explanations of isolated events. Scientific theories generally are about closed systems governed by laws of succession,[37] and the explanations they afford are of classes of phenomena. And to these, laws of forms (*) are utterly irrelevant. Again aside from theories used in the design of measuring apparatus or special-case versions of more global theories, theories with laws of quasi-succession are relatively rare—although one of our most impressive global theories (the genetic theory of natural selection) fundamentally employs such laws. But, as we saw in the previous subsection, such theories can yield explanations beyond $t + 1$ only when augmented by boundary conditions that make them act as if they had laws of succession. Further, since laws such as those of forms (*) ignore the crucial importance of internal substates in laws of quasi-succession, they are theoretically impoverished and highly unrepresentative. By ignoring the rich structural properties of theories, earlier accounts of explanation have missed much of what is most central to the ability of theories to give explanations.

The Problem of Laws of Coexistence

If one carefully peruses the various counterexamples posed by Bromberger, Salmon, and some others to Hempel's D-N model of explanation,[38] one finds that they frequently involve "laws of coexistence," of which the "ideal gas laws," such as

$$(**) \qquad\qquad PV \;=\; nRT,$$

are typical, and which, as Salmon notes, "are non-causal."[39] The flagpole case we considered in the first section is generally presented as employing a law of coexistence. These laws, as commonly stated, are *not* state-transition laws. They merely assert that such equilibria will hold—"come what may," without any specification of *how* alterations in any of the state parameters may occur. In a certain sense they are *degenerate* statistical laws of succession in that they say "any equilibrium state as defined by an equation such as $(**)$ is equi-probable."[40] But as actually formulated they say nothing explicitly about state transitions. However, it is not difficult to reformulate such laws to be explicitly state-transitional—to have the following form (for nonstatistical laws):

> *Definition*: Let t, t', and t'' be variable ranging over an α-time sequence. Then a nonstatistical law of coexistence is a relation $R[s(t),\; s'(t')]$ meeting the following conditions for any t, t', t'' such that $t \;<\; t'$ and $t' \;<\; t''$:
> for every s, $<s(t),\; s(t')> \; \varepsilon \; R$;
> for every s, s', if $<s(t),\; s'(t')> \; \varepsilon \; R$,
>   then $<s'(t'),\; s(t)> \; \varepsilon \; R$;
> for every s, s', and s'', if $<s(t),\; s'(t')> \; \varepsilon \; R$ and
> $<s'(t'),\; s''(t'')> \; \varepsilon \; R$, then $<s(t),\; s''(t'')> \; \varepsilon \; R$.

So reformulated, if such laws are empirically true, I have no qualms about their being the basis for "how could" or (if augmented) "how did" explanations. Still, one should be worried—since they are at the mercy of factors not reflected in the law for state transitions and thus ultimately are quite nondeterministic: these are the ultimately indiscriminate nondeterministic laws. And they are explanatorily tepid. Still such a reformulation minimally gives a weak "how could" explanation. With respect to explanations *why* it is totally impotent; for such laws never can meet the unique-terminus requirement.

Still, it might seem that I have allowed too much. For consider the notorious flagpole counterexample to Hempel's D-N model, wherein laws of coexistence allow the shadow cast by a flagpole to "explain" the angle of the sun to the horizon. Do I not allow such cases? No! The most I allow is a degenerate explanation of how the collective angles, height of the flagpole, and length of the flagpole shadow do lawfully covary—without any imputation of cause. The troublesome causal-influence abnormalities occur only when one illegitimately attempts to convert laws of coexistence into laws of succession or quasi-succession; and to do so is radically to transform the laws in ways that do violence to the time variables and temporal asymmetry. For example, in the flagpole counterexample, when stated in a form analogous to (**)—as it generally is stated in the explanation literature—the length of the shadow, the height of the flagpole, and the angle of incidence of the sun to the ground appear to be simultaneous variables. But when the flagpole case is formulated with an appropriately discriminative time scale and a fixed height for the flagpole, the angle of incidence of the sun at t determines the length of the shadow at time $t + \varepsilon$. In short, we have a law that fails to be a type (**) law of coexistence—rather it is a deterministic law of quasi-succession that incorporates precisely the sorts of temporal asymmetries needed to block the standard counterexamples pertaining to explaining *why*. Once this clarifying move is made, I believe, not only can we explain away such troublesome cases but we can also, depending upon the form of the laws, accommodate them to the accounts of theoretical explanation given above.

The foregoing prompts me to advance the following hypothesis: Insofar as we confine our attention to explaining subsequent complete system states as functions of prior ones, and insofar as such state-transitions are mediated by empirically true theories, laws of coexistence raise no special problems for "how could" explanations. They afford no "why" explanations since they cannot meet the unique-terminus requirement for "why" explanations introduced variously above. Standard counterexamples are pseudo-counterexamples—explainable away by adequate attention to details. This is especially true of the standard "counterexamples" involving temporal asymmetry or those that attempt to push theories beyond their explanatory capabilities—which capabilities have, I hope, been better articulated in this paper.

Conclusions

I began this paper by trying to motivate the idea that a nondeductivist approach to analyzing scientific explanations is more promising than the deductivist approaches that have tended to dominate philosophical discussions for the last half-century. In pushing a particular approach to a philosophical problem, there comes a point at which the only plausible way to convince people is to carry through—to produce the sort of analysis advocated in fine detail and show that it can resolve difficulties that have foundered the analyses associated with competing approaches. I have accordingly devoted the bulk of the present paper to developing a specific nondeductivist analysis and then showing in detail how it can avoid not only the problems deductivist approaches have succumbed to but also those that Salmon's nondeductivist S-R attempts have encountered. The resulting analysis is nondeductivist in precisely the way the semantic conception is: both can be deployed in deductive deliberations and arguments, but such deductive moves are not integral parts of either analysis.

I do not begin to think I have solved all the philosophical problems of theoretical explanation, to say nothing of scientific explanation in general. I hope only to have mounted an enticing case for the claim that a nondeductivist approach to explanation is the most promising option in the present philosophical context. Central to the case is the idea that treating the structure of theories seriously is a productive way to rethink mainstream explanatory issues in philosophy of science.

One clarifying caveat concerning the nondeductivist analysis presented here: For all my discussion I have not offered a "standard" philosophical analysis of "theoretical explanation." Specifically, I have not offered *necessary* or *sufficient* conditions for something's being a *theoretical explanation*, although I have given necessary conditions for various kinds of laws' being able to yield explanations *how could, how did,* and *why*. Since I cannot prove that the sorts of laws I have considered are exhaustive of the laws that can figure in theories,[41] it is premature for me to claim to have found necessary conditions for something's being a theoretical explanation. As to sufficient conditions, I doubt whether any exhaustive set of them will or can be forthcoming. For any conclusive account of what a theoretical explanation is must be parasitic upon an account of what it is to be a *scientific* theory. While I believe the semantic conception provides a defensible account of what it is to be *a theory,* I do not believe it is, or potentially can become, an adequate account of what a *scientific* theory is. Whether a theory is scientific ultimately depends upon the "domain" the science deals with and its evolving standards as to what is scientific, and thus ultimately is not a matter for philosophical fiat or decision. However, one can present, as I have here, a philosophical account of what it is about theories that *enables* them to provide scientific explanations, and one can explain how and what sorts of explanations they afford. I believe doing so can be philosophically and scientifically illuminating. But I do not believe that I or any other philosopher of science can have the last word on the subject.

Notes

1. Wesley Salmon, "Theoretical Explanation," in S. Körner, ed., *Explanation* (Oxford: Blackwell, 1975), pp. 119–120.

2. I have investigated these in a number of papers concerned with the semantic conception of theories: "The Meaning and Use of Models in Mathematics and the Exact Sciences" (Ph.D. dissertation, University of Michigan, 1967); "What's Wrong with the Received View on the Structure of Scientific Theories?" *Philosophy of Science* 39 (1972):1–19; "Theories, Their Formulations, and the Operational Imperative," *Synthese* 25 (1973):129–164; "Some Philosophical Problems in Biological Speciation and Taxonomy," In J. Wojciechowski, ed., *Conceptual Basis of the Classification of Knowledge* (Munich: Verlag Dokumentation, 1974), pp. 190–243; "Theoretical Laws," in M. Przełecki et al., eds., *Formal Methods of the Methodology of Social Science* (Wrocław, Poland: Ossolineum, 1976), pp. 247–267; and "The Search for Philosophic Understanding of Scientific Theories," in Frederick Suppe, ed., *The Structure of Scientific Theories*, 2d ed. (Urbana: University of Illinois Press, 1977), pp. 3–241, esp. pp. 221–230. All but the first and last works have been incorporated into my *The Semantic Conception of Theories and Scientific Realism* (Urbana: University of Illinois Press, forthcoming).

3. See Sylvain Bromberger, "Why Questions," in R. Colodny, ed., *Mind and Cosmos: Explorations in the Philosophy of Science* (Pittsburgh,

Pa.: University of Pittsburgh Press, 1966), pp. 86–111, and Stephen Toulmin, *Foresight and Understanding* (London: Hutchinson, 1963).

4. The most important are Carl G. Hempel and Paul Oppenheim, "Studies in the Logic of Explanation," *Philosophy of Science* 15 (1948):135–175, and Carl G. Hempel, "Aspects of Scientific Explanation," in his *Aspects of Scientific Explanation and Other Essays in the Philosophy of Science* (New York: The Free Press, 1965), pp. 331–496.

5. Bromberger, op. cit., and Wesley Salmon et al., *Statistical Explanation and Statistical Relevance* (Pittsburgh, Pa.: University of Pittsburgh Press, 1971).

6. Salmon, in Salmon et al., op. cit.; and Richard Jeffrey, "Statistical Explanation vs. Statistical Inference," in Salmon et al., op. cit., pp. 19–28.

7. A number of other counterexamples have been raised in the literature by Michael Scriven, Toulmin, and others. Hempel has responded to these in his "Aspects of Scientific Explanation." I do not find Hempel's rejoinders there as convincing as the ones I consider here. I have discussed some of them in "Afterword—1977," in my *Structure*, pp. 617–730, esp. pp. 619–623.

8. E.g., in a lecture at Johns Hopkins University in spring 1974.

9. In the works cited in note 6 above.

10. Bromberger, op. cit.

11. Toulmin, op. cit.

12. John Meixner, "Homogeneity and Explanatory Depth," *Philosophy of Science* 46 (1979):372 n. 14. Meixner points out that this rule is superfluous whenever the reference classes are homogeneous (as Salmon's analysis requires).

13. Meixner, ibid.

14. Even if $B \quad C$ and $A \quad C$ weren't homogeneous reference classes, that wouldn't affect things, for we could break them down into such classes. But that wouldn't alter things since the event we're concerned with belongs to AC and to BC.

15. Merrilee Salmon and Wesley Salmon, "Alternative Models of Explanation," *American Anthropologist* 81 (1979):69–74.

16. Wesley Salmon, "Theoretical Explanation"; "Why Ask, 'Why'? An Inquiry Concerning Scientific Explanation," *Proceedings and Addresses of the American Philosophical Association* 51 (1978):683–705; and "An 'At-At' Theory of Causal Influence," *Philosophy of Science* 44 (1977):215–224. A more comprehensive treatment, refinement, and development of these ideas is found in his *Scientific Explanation and the Causal Structure of the World* (Princeton, N.J.: Princeton University Press, 1984). His paper in the present volume sketches the basic ideas.

17. L. Jonathan Cohen, "Comment [on Salmon's 'Theoretical Explanation']," in Körner, op. cit., pp. 152–159.

18. Hempel, "Aspects of Scientific Explanation," p. 334; some italics are mine.

19. See, e.g., David Harrah, *Communication: A Logical Model* (Cambridge, Mass.: MIT Press, 1963); Nuel D. Belnap, Jr., *An Analysis of Questions: Preliminary Report* (Santa Monica, Calif.: Systems Development Corporation, 1963); and Nuel D. Belnap, Jr., and Thomas B. Steel, *The Logic of Questions and Answers* (New Haven, Conn.: Yale University Press, 1976). Bromberger, op. cit., also is relevant.

20. See Bromberger, op. cit., and Belnap and Steel, op. cit., for related discussion. At present there is no satisfactory treatment of the relationships between answers to "why" and "how" questions and their presuppositions.

21. For more details, see the works cited in note 2 above.

22. In statistical theories, not only are possible paths specified but so are the probabilities of allowed state-transitions. The resulting complications (e.g., for empirical truth conditions) will be omitted in the remainder of this sketch of the semantic conception.

23. For further discussion of conceptual devices, see my "Afterword—1977," op. cit., esp. pp. 706–716.

24. See my "Theoretical Laws," pp. 258–259, for the arguments that this is so.

25. This example is adapted from John Kemeny et al., *Finite Mathematical Structures* (Englewood Cliffs, N.J.: Prentice Hall, 1959), pp. 416–417. That the example is an oversimplification and not really an empirically true theory does not affect my argument.

26. I will have more to say about this when we consider the problem of laws of coexistence below. For the moment it will suffice to note that the state-transition paths are time-directed graphs.

27. I am assuming here, for simplicity of illustration, that time is discrete.

28. Another important sort, laws of coexistence, will be discussed below in the subsection "The Problem of Laws of Coexistence."

29. In my "Theoretical Laws," I first identified laws of quasi-succession and gave defenses of the various claims about them made here. That initial characterization of laws of quasi-succession was defective, however, and the present version replaces the definitions given there. I am grateful to Ronald Giere for advice leading to the present formulation.

30. See my "Theoretical Laws," Sec. IV, for the details.

31. In its standard formulation this theory has a deterministic law of quasi-succession with statistical state parameters (probabilistic population distribution functions). Presumably (and with great complexity) it could be presented as a theory with a statistical law of quasi-succession. That it is not probably reflects the facts that (a) every theory with statistical laws can be reformulated equivalently as a theory with a deterministic law over

probabilistic states, and (b) the latter tend to be more compact, manage-able, and mathematically tractable. See my "Meaning and Use of Models," pp. 144–149.

32. I intend this formulation to allow the possibility that if only partial knowledge of the external state history enables one to determine a unique terminous, an explanation why can be given.

33. Salmon's S-R model will sanction such explanations. Hempel's I-S model would not allow them since the probabilities are too low.

34. A theory must have a notion of state, and the law must be one of state-transition. Given the data the American Cancer Society collected, states would have to be something like the following:
<years smoked, amount smoked per day, age, cancer—yes or no>.
Then, given some appropriate time scale (say, year intervals), one would have to have a law specifying state-transitions (via a statistical law either of succession or of quasi-succession). The actual data of the study (or all other cancer studies for that matter) are not adequate to specify or even validate such a law. For a description and philosophical analysis of the study, see Ronald Giere, *Understanding Scientific Reasoning* (New York: Holt, Rinehart and Winston, 1979), secs. 12.3 and 14.5.

35. For discussions of statistical notions of causality, Giere, see ibid., chap. 12; Ronald Giere, "Causal Systems and Statistical Hypotheses," in L. J. Cohen and M. B. Hesse, eds., *Applications of Inductive Logic* (New York: Oxford University Press, 1980), pp. 251–270; and Patrick Suppes, *A Probabilistic Theory of Causality*, vol. 24 of *Acta Philosophia Fennica* (Amsterdam: North Holland, 1970).

36. E.g., I believe this failure plays an important role in the inconclusive controversy between Scriven and Hempel over the adequacy of the D-N model.

37. The complications due to laws of coexistence will be considered in the next subsection.

38. See the works cited in notes 3 and 6 above.

39. Salmon, "Why Ask 'Why?'?" p. 687.

40. This is the sense of "law of coexistence" I defined in "Theoretical Laws," following van Fraassen.

41. See my "Theoretical Laws" for discussion of and conjectures on this issue.

Contributors

Ronald N. Giere is professor of philosophy and director of the Minnesota Center for Philosophy of Science at the University of Minnesota. He is the author of *Understanding Scientific Reasoning* and *Explaining Science: A Cognitive Approach.*

Adolf Grünbaum is Andrew Mellon Professor of Philosophy at the University of Pittsburgh. His books include *Philosophical Problems of Space and Time* and *The Foundations of Psychoanalysis.* His work has focused on the philosophy of physics, the theory of scientific rationality, and the philosophy of psychiatry.

Carl G. Hempel is professor of philosophy, emeritus, at Princeton University and the University of Pittsburgh. His research and writings have been concerned principally with logical and methodological aspects of inquiry in the natural and the social sciences, and in historiography.

Henry E. Kyburg, Jr. is Burbank Professor of Moral and Intellectual Philosophy and professor of Computer Science at the University of Rochester. His main interests are in the areas of the logical foundations of statistical inference, theory and measurement, and epistemology and inference.

Wesley C. Salmon is University Professor of Philosophy at the University of Pittsburgh. Among his books are *Logic, Foundations of Scientific Inference,* and *Scientific Explanation and the Causal Structure of the World.* His main interests are in scientific explanation, causality, and confirmation.

Frederick Suppe is professor of philosophy at the University of Maryland. He is the editor of *The Structure of Scientific Theories* and has just completed another book, *The Semantic Conception of Theories and Scientific Realism.*

Designer: U.C. Press Staff
Compositor: Janet Sheila Brown
Text: 10/12 Sabon
Display: Sabon
Printer: Braun-Brumfield
Binder: Braun-Brumfield